DEMOLITION AND REUSE OF CONCRETE AND MASONRY

Publisher's Note

This book has been produced from camera ready copy provided by the individual contributors.

This method of production has allowed us to supply finished copies to the conference delegates in advance of the meeting.

DEMOLITION AND REUSE OF CONCRETE AND MASONRY

VOLUME TWO

Reuse of Demolition Waste

Proceedings of the Second International Symposium held by RILEM (the International Union of Testing and Research Laboratories for Materials and Structures) organized by the Building Research Institute, Ministry of Construction, Japan and co-organized by Nihon University, Japan

Nihon Daigaku Kaikan
Tokyo, Japan
November 7–11, 1988

EDITED BY
Y. Kasai

Taylor & Francis
Taylor & Francis Group
LONDON AND NEW YORK

First published in 1988 by
Taylor and Francis,
2 Park Square, Milton Park, Abingdon,Oxon, OX14 4RN
52 Vanderbilt Avenue, New York, NY 10017

First issued in paperback 2020

Taylor & Francis is an imprint of the Taylor & Francis Group, an informa business

British Library Cataloguing in Publication Data

Demolition and reuse of concrete and masonry
1. Concrete structures. Demolition
I. Kasai, Y.
624'.1834

ISBN 0 412 32110 6
ISBN 0 412 34480 7 V.1
ISBN 0 412 34490 4 V.2

ISBN 13: 978-0-412-32110-8 (set)
ISBN 13: 978-0-367-65946-2 (pbk)
ISBN 13: 978-0-412-34490-9 (hbk)

ORGANIZING COMMITTEE

Dr S. Fujimatsu	*General Director, Building Research Institute, Ministry of Construction, (Chairperson), Japan*
Professor A. Enami	*Nihon University, Japan*
Professor K. Kamimura	*University of Utsunomiya, Japan*
Professor Y. Kasai	*Nihon University, Japan*
Mr K. Kato	*Tokyo Electric Power Company, Japan*
Professor K. Kishitani	*Nihon University, Japan*
Mr S. Koizumi	*Building Research Institute, Ministry of Construction, Japan*
Professor S. Nagataki	*Tokyo Institute of Technology, Japan*
Mr N. Sato	*Ministry of Construction, Japan*
Mr T. Tanimoto	*Public Works Research Institute, Ministry of Construction, Japan*
Professor F. Tomosawa	*University of Tokyo, Japan*
Mr M. Yokota	*Japan Atomic Energy Research Institute, Japan*
Professor T. C. Hansen	*Technical University of Denmark, Denmark*
Mr M. Fickelson	*General Secretariat, RILEM, France*

NATIONAL EXECUTIVE COMMITTEE

Professor K. Kamimura	*University of Utsunomiya (Chairperson), Japan*

INTERNATIONAL SCIENTIFIC COMMITTEE

Professor Y. Kasai	*Nihon University, (Chairperson), Japan*
Mr T. Egashira	*Kaihatu Denki Co., Ltd, Japan*
Mr T. Fujii	*Shimizu Corporation, Japan*
Professor T. Fukuchi	*Nihon University, Japan*
Professor T.C. Hansen	*Technical University of Denmark, Denmark*
Dr C.F. Hendriks	*Road Engineering Division, Rijkswaterstaat, Netherlands*

Dr H. Kaga	*Taisei Corporation, Japan*
Dr T. Kemi	*Institute of Technology, Toda Construction Co., Ltd, Japan*
Dr K. Kleiser	*University of Karlsruhe, West Germany*
Mr S. Kobayashi	*Public Works Research Institute, Ministry of Construction, Japan*
Mr E.K. Lauritzen	*Demex, Denmark*
Dr P. Lindsell	*University of Oxford, England*
Professor Y. Malier	*Laboratoire Central des Ponts et Chaussees, France*
Dr Y. Masuda	*Building Research Institute, Ministry of Construction, Japan*
Mr M.T. Mills	*Institute of Demolition Engineering, Griffiths-McGee Demolition Co., Ltd, England*
Mr P. Mohr	*A/S Skaninavisk Spendbeton, Denmark*
Dr C. Molin	*Statens Provningsanstalt (National Testing Institute), Sweden*
Professor T. Mukai	*Meiji University, Japan*
Dr M. Mulheron	*University of Surrey, England*
Professor S. Nagataki	*Tokyo Institute of Technology, Japan*
Professor N. Nishizawa	*Chuou University, Japan*
Mr C. de Pauw	*Centre Scientifique et Technique de la Construction, Belgium*
Dr G. Ray	*Concrete Paving Consultant, USA*
Mr E. Rousseau	*Centre Scientifique et Technique de la Construction, Belgium*
Professor M. Sakuta	*Nihon University, Japan*
Dr R. Schulz	*Institute fur Baustoffprufung, West Germany*
Professor T. Soshiroda	*Shibaura Institute of Technology, Japan*
Mr Y. Takahashi	*Building Research Institute, Ministry of Construction, Japan*
Professor F. Tomosawa	*University of Tokyo, Japan*
Professor K. Torigai	*Science University of Tokyo, Japan*

Contents

VOLUME TWO REUSE OF DEMOLITION WASTE

Preface

RILEM Technical Committee 37 DRC on Demolition and Reuse of Concrete was formed in 1976 and held its first meeting at the Building Research Station in Garston (UK) in June of 1977 under the chairmanship of Dr L.H. Everett. In 1978 the first RILEM TC 37–DRC state-of-the-art report was published on recycled concrete as an aggregate for concrete [1].

After the committee was reorganized in 1981 and the author of this preface became chairman, a second committee meeting was held in Copenhagen (DK) in December 1982. Since then the committee has held yearly meetings in the Netherlands, England, Belgium, France and Japan.

The following general terms of reference of the committee were agreed on at the meeting in Copenhagen in 1982.

1. To study the demolition techniques used for plain, reinforced, and prestressed concrete and to consider developments in techniques.
2. To study technical aspects associated with reuse of concrete and to consider economical, social and environmental aspects of demolition techniques and reuse of concrete.

Three task forces were formed, each with its own specific terms of reference.

Task Force 1 surveyed, on the basis of the existing literature, methods of demolition and fragmentation including economic, social and environmental aspects. It published its findings in two state-of-the-art reports, one on demolition techniques in general [2] and another on blasting of concrete [3].

Task Force 2 collected and surveyed codes and regulations concerning demolition in various countries. It did not issue a separate state-of-the-art report. Instead its findings were included in [2,3,4,5].

Task Force 3 studied technical aspects associated with reuse of concrete and considered economic, social and environmental factors. It published its findings in two state-of-the-art reports, one on the reuse of concrete as concrete aggregates [4] and another report on the reuse of mixed concrete and masonry rubble as aggregate for concrete [5].

The committee arranged the first international symposium on demolition and recycling of concrete in Rotterdam in 1985 in co-operation with the European Demolition Association (EDA). The symposium proceedings were published in [6] and [7]. The symposium gave valuable input to the work of the committee from an industrial point of view. Developments were fast, and it was soon decided to hold a second international RILEM symposium on demolition and reuse of concrete already in 1988 in Tokyo in order once more to make it possible for people from science and practice and from all over the world to communicate and exchange experience before the committee is dissolved at a final meeting in Tokyo in 1988. As chairman of RILEM TC 37–DRC I sincerely hope the symposium will be successful and that the proceedings will serve as a source of inspiration for further research and development in the fields of demolition and reuse of concrete.

Moreover, I want to thank the following persons who have served as members and corresponding members of the committee over the years. Members Mr R.C. Basart (NL), Dr Ch.F. Hendriks (NL), Professor P. Lindsell (GB), Professor Y. Kasai (Japan), Dr K. Kleiser (D), Dr R.R. Schulz (D), Professor Y. Malier (F), Mr R. Hartland (GB), Mr T.R. Mills (GB), Mr P. Mohr (DK), Dr C. Molin (S), Mr G. Ray (USA), Mr C. de Pauw (B), Mr E. Rousseau (B), Mr E.K. Lauritzen (DK), Secretary from 1982–1985, and Dr M. Mulheron (GB), Secretary from 1985–1988. Corresponding members: Mr F.D. Beresford (AUS), Mr M. Whelan (AUS), Mr A.D. Buck (USA), Dr S. Frondistou-Yannas (USA), Mr J.M. Loizeaux (USA), Mr J.F. Lamond (USA). Our very special thanks go to the European Demolition Association for its loyal co-operation in the work of the committee, and to the Japanese Building Research Institute for organizing this symposium.

Torben C. Hansen

LIST OF REPORTS ISSUED BY RILEM TECHNICAL COMMITTEE 37–DRC

[1] Nixon, P.J. (1978) Recycled concrete as an aggregate for concrete – a review. *Materials and Structures*, **11** (65) September–October, pp. 371–8.

[2] Task Force 1 – RILEM Technical Committee 37–DRC (1985) *Demolition Techniques*. European Demolition Association, Wassenaarseweg 80, 2596 CZ Den Haag, The Netherlands, Special Technical Publication, May.

[3] Molin C. and Lauritzen, E. (1988) *Blasting of Concrete. Localized Cutting and Partial Demolition of Concrete*. Special Report 09, National Testing Institute, Box 5608, 11486 Stockholm, Sweden.

[4] Hansen, T.C. (1986) Recycled aggregates and recycled concrete aggregate. Second state-of-the-art report, developments 1945–1985. *Materials and Structures*, **19** (111), May–June, pp. 201–46.

[5] Hendriks, Ch.F. and Schulz, R.R. Recycled masonry rubble as an aggregate for concrete. State-of-the-art report, developments 1945–1985. *Materials and Structures* (in press).

[6] EDA–RILEM (1985) *Demolition Techniques. Proc. First International EDA–RILEM Conference on Demolition and Reuse of Concrete*, Rotterdam, 1–3 June, **1**, European Demolition Association, Wassenaarseweg 80, 2596 CZ Den Haag, The Netherlands.

[7] EDA–RILEM (1985) *Reuse of Concrete and Brick Materials. Proc. First International EDA–RILEM Conference on Demolition and Reuse of Concrete*, Rotterdam, 1–3 June, **2**, European Demolition Association, Wassenaarseweg 80, 2596 CZ Den Haag, The Netherlands.

Foreword

The buildings and the urban areas where we live have had to change in harmony with the needs of the age. In these circumstances, it is natural to consider demolition work as being as important as construction work in the sound development of urban areas. Demolition work is inevitable for the revitalization of urban areas. In congested cities, in particular, the quality of the demolition technique becomes an essential element which determines the success of the revitalization of the city. In addition to efficiency in demolition, strategies must be adopted to avoid noise, vibration and dust which affect the surrounding environment and there must be efficient disposal of the waste products. The effective reuse of these waste products can be of great importance socially as well as economically.

As far as Japan is concerned, concrete and masonry construction has only a one hundred year history. Since the great earthquake in the Kanto region, reinforced concrete has been used for the main, large buildings. Soon after the second world war, many buildings of poor quality were constructed because of the need for low-cost buildings and they remain even now. To date, many buildings with reinforced concrete have been demolished and new buildings which satisfy modern requirements have been constructed in their place.

In this symposium, various demolition techniques which have been accumulated and tested in recent years will be presented. They will be helpful for demolition work all over the world. Some presentations will report research carried out on demolition work at atomic power stations or at buildings where asbestos fibres were used as building materials. I believe that they will meet our expectations.

I would like to thank all the members of RILEM TC37 for their honest efforts to undertake research on demolition techniques over a long period. I would also like to extend my thanks to the members of the International Scientific Committee who have worked hard to hold this symposium. I hope it will be useful in efficiently establishing the cities of tomorrow.

Dr Susumu Fukimatsu

Introduction

At this time when the second RILEM Symposium on the Demolition and Reuse of Concrete and Masonry is to be held in Japan, I would briefly like to describe the history of research and development in this field in Japan. In the near future the vibrationless and noiseless methods of demolition developed for the congested urban areas found in Japan will be utilized for demolition work in cities around the world.

As far as concrete structures are concerned the mechanisation of demolition work started in the late 1950s with the introduction of pneumatic hand hammer breakers and steel balls. Using these tools the slabs of multi-storey buildings were first broken up by 'balling' and then the beam ends were crushed by hand hammer breakers. Finally the remaining large walls, often of several spans and storeys, were then felled in a single operation. However, this method was dangerous and resulted in a number of accidents.

Due to the high economic growth policy and changes in peoples' attitudes, there were frequent complaints and claims made against demolition activities in congested cities. As a result the construction industry was obliged to develop safer, quieter and less intrusive methods of demolition.

The research committee on 'Demolition and Removal Methods' was started in these days and the first technical book in this field *Demolition Method for Concrete Structures* was published in 1970. In 1971 the Building Contractors Society (BCS) started the 'Committee on Demolition of Reinforced Concrete Structures' which drew its members from the research and engineering staff of major contractors along with academics from various universities. This committee conducted research and development work into a variety of demolition techniques such as jacking, explosives, and rebar heating methods by either direct or induced current. In 1978 the committee published the *Standard of Public Nuisanceless Demolition Method of Reinforced Concrete Structures* and in 1987 also published a *Recommended Proposal of Demolition Method of Underground Reinforced Concrete Structures*.

During this period, general contractors and manufacturers of construction plant joined together in developing hydraulic 'C-shaped' concrete crushers, diamond cutters and flame jetting methods. A most important development was the introduction of concrete crushers from England in about 1975. Stimulated by this, and using the experience gained from the production of 'C-shaped' crushers a number of highly efficient concrete crushers were developed and these can be seen in use to this day. The development of chemical expansive demolition agents, started in 1967, was a solely Japanese initiative. They were available as commercial products by 1978. Developments have continued in this area and today it is possible to purchase materials capable of crushing concrete within one hour of application.

The study of the demolition of the Japan Power Demonstration Reactor (JPDR) by the Japan Atomic Energy Research Institute (JAERI), which started in about 1979, resulted in many useful developments for the demolition of reinforced concrete structures by explosives, large diamond cutters, core boring machines and abrasive water jetting. In addition methods for stripping surface concrete by the use of microwaves were investigated. The removal of surface concrete by the rebar heating method using alternating current was another Japanese development and after the initial experimental trials were complete in 1968, it was later used for the demolition of special structures. In 1981, diamond wire saw for cutting reinforced concrete was introduced and it is expected that this method will be the subject of further developments over the coming year.

Study of the reuse of concrete waste in Japan started in about 1971. In 1974 the BCS formed the 'Committee on Disposal and Reuse of Construction Waste'. This committee conducted many successful experiments on the production of recycled concrete aggregate, the study of recycled concrete, chippings from waste wood and the production of wood chip concrete. In 1977 this body published the *Proposed Standard for the Use of Recycled Aggregate and Recycled Aggregate Concrete*. Later on, during the period 1981–1985, the Ministry of Construction conducted a study to encourage the reuse of construction waste for new construction work and introduced a standard for the reuse of demolished concrete and waste timber.

In this way the methods of demolishing concrete structures have developed rapidly in Japan in order to meet the strict requirements for demolition methods imposed by its citizens. The successful development of these methods is the result of the efforts of the demolition related industries, academic institutions and public authorities who have joined forces to ensure that these requirements are met. Indeed the reason for holding the Symposium in Tokyo is that the RILEM Committee 37–DRC considered that the development of demolition methods and use of modern demolition techniques was well advanced in Japan.

We sincerely expect that this Symposium will contribute to the development of Demolition and Reuse of Concrete and Masonry Structures.

Yoshio Kasai
Chairman of Scientific Committee

CONCRETE WITH RECYCLED RUBBLE - DEVELOPMENTS IN WEST GERMANY

R.-R. SCHULZ
Institute for testing materials, Waldkirch, West Germany

Abstract
Except for burnt clay bricks, processed rubble cannot be re-used as concrete aggregate according to German standards. In 1987 a pilot project was started in West Berlin in order to get a single permission for re-using about 5000 tons of processed rubble. Several aptitude tests had to be performed in order to find out whether the products of the recycling plants in West Berlin can be used for this purpose. The test results were compared with data from literature. The mechanical properties of concrete with recycled aggregate seem to be sufficient for many purposes. Problems may occur because of the reduced resistance against frost and carbonation.
Key words: Recycling, Aggregate, Concrete, Masonry rubble, War debris

1 Introduction

The development of recycling technology in Germany dates back for about 1900 years. At that time the Romans had occupied large areas of Europe and also in the river Rhine valley they built fortification walls, roads, aqueducts and the like with concrete using rock material but also crushed burnt clay brick as concrete aggregate. More information and even experimental data can be found in /1/. About forty years ago recycling brick became most important again due to the huge amount of war debris. Until 1955 about 11.5 million cubic metres of brick were recycled as concrete aggregate in West Germany. In the year 1951 a standard for concrete with crushed brick was introduced /2/. This standard was withdrawn respectively transfered into the standard for lightweight aggregate /3/, when the amount of debris was falling short. Crushed burnt clay brick may still be re-used as lightweight aggregate after DIN 4226 /3/.

Unfortunately other material to be recycled like concrete or masonry rubble which today consists of more or less different building materials cannot be re-used as concrete aggregate without single or general permissions by the building authorities. As it is rather complicated and expensive to get such an approval, private companies were prevented from making the first move in that direction. On the other hand there will be no standard for recycled concrete or recycled masonry if there is not enough experience acquired through the producing and re-using these materials under the control of the authorities.

This sounds discouraging though recycling technology is often proclaimed to be necessary from the viewpoint of environmental politics. In fact, the government is increasing its support. Recently a research work /4/ was financed by the Federal Ministry for Construction, Town and Country Planning. This study was to show the chances of this technology and to provide technical instructions and recommendations for employing this technology. For this purpose all available data on recycling building rubble as concrete aggregate had to be analysed in order to find correlations of practical use. One of the most important features was to consider the literature on the post-war recycling activities in West Germany /4, 5/.

Another recent research work /6/ was to find out whether the products of a recycling plant in Düsseldorf which applies wet processing for cleaning up the upgraded material are convenient to be used as an aggregate for concrete. The tests included large-scale specimens. These features are very interesting from an scientific point of view but it seems that laboratory studies alone will not be sufficient. Large scale practical initiatives are necessary to show the environmental, technical and economical prospects of the re-using of building rubble as an aggregate for concrete.

Since West Berlin is cut off from natural raw material supplies all aggregate has to be transported on a long distance from West Germany through the German Demokratic Republic or otherwise has to be imported from the GDR. This is why the building senator of West Berlin is very much interested in new methods that can help solve the problems with missing raw material supplies and disposal areas. He took the initiative to establish a pilot project for re-using concrete and masonry on a large scale at the end of 1987 in order to get a single permission for the re-used aggregate. First about 5000 tons of processed rubble from already existing plants were to be used as concrete aggregate for a foundation wall of a new recycling plant to be installed. Then, after the plant with a modern and efficient technical supplement has started work it is scheduled to get a general permission for its products. This recycling plant is planned to be capable of producing concrete aggregate in order to become more independent of transporting and importing.

2 The constituents of the rubble

Only in few cases, for example if concrete road pavements and airport runways are rebuilt, the rubble consists of nearly 'clean' and constant material. But more often one can find masonry bricks, masonry mortar, screed material, plaster roof tiles and other ceramical products in the rubble. Moreover rubble even contains metals, asphalt, wood, plastics, glass, gipsum, paper and insulation material which have to be regarded as contaminants.

The more the disposal areas decrease and disposal fees increase the chances of interaction, preselection and the predetermination of the constituents on site will improve. This means the kind and the constituents of the rubble to be recycled might up to a certain degree be regulated by the disposal fee. The better the incoming rubble is the lower the disposal fee could be. Valuable building rubble is: cementous building material like concrete, but also natural stone (even already used gravel and sand), masonry rubble and asphalt. Until now most of these materials are re-used for road-building purposes or as refill for excavated areas. But alternatively with the exception of asphalt most of these materials could be re-used as concrete aggregate.

3 Recycling plants

In West Germany there are more than 100 recycling plants. Most of them are smaller ones with installations just for crushing and screening the more or less preselected recycled rubble. Compared with the USA more impact crushers without secondary crushing are used. These simple plants are not capable of removing contaminants (with the exception of iron and steel by self cleaning magnets and rubble fines by screening). Only a few larger plants (probably 10 %) in the more populated areas of this country apply washing or air sifting procedures for removing lightweight particles such as dirt, clay lumps, wood, paper, plastics, textiles etc. in order to make their products fit for more exacting purposes such as antifreeze layers or base courses of roads which may justify higher prices.

4 Properties of processed rubble

4.1 Kind and constituents

By limiting the contents of different constituents the prediction of properties will up to a certain degree be rendered possible. According to the recommendations in /7/, recycled concrete must contain more than 95 % by weight pure concrete with a bulk density over 2100 kg/m^3 and recycled masonry which must consist of at least 65 % by weight concrete and/or burnt clay brick and/or silica limestone. There should be similar regulations for recycled burnt clay brick etc..

Not only will quality control tests be expensive and time consuming, severe difficulties are expected to occur in defining the material and in producing recycled aggregate of constant properties. How could this be achieved since everybody knows that waste is really inhomogenous? The following tries to discuss what has to be done in order to define the aggregate and to meet the quality requirements.

There are two major tasks for which the best results must be achieved. One of them is to equalize the product in order to fit an acceptable range of properties on an acceptable level (recycled masonry rubble). The other is to extract special parts of the waste in order to fit a given level of properties within a given range (for example pure burnt clay bricks or pure concrete). The first task may be achieved by equalizing procedures the second especially by preselecting and sorting. Preselecting may take place on demolishing sites or at the recycling plant by directing the trucks to different stockpiles or by feeding the crusher with for example pure concrete after visual criteria. Equalizing can only be achieved by stockpiling larger amounts of material and mixing it. At least the daily output has to be stockpiled. Mixing may be done by blending beds or by shifting the material from one stockpile to the other for example by wheel dozers. The quality and uniformity of the equalized material can be tested by quick-tests like heaped bulk density control which is discussed later on. At least all necessary tests required for natural gravel or lightweight aggregate should be performed.

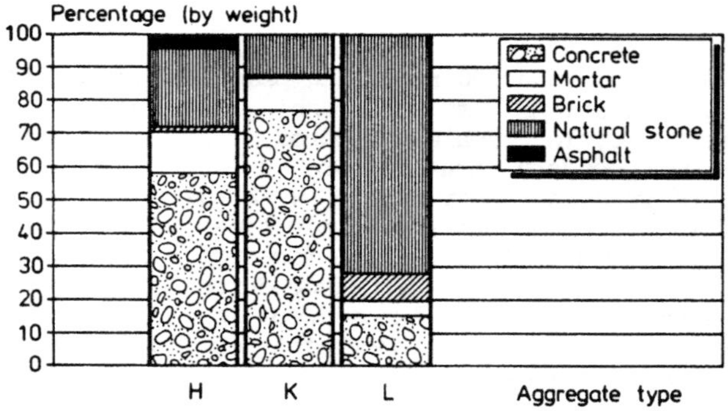

Fig. 1: Main constituents of three recycled aggregates

The composition of three coarse aggregate types from West Berlin are shown in figure 1. Since the samples were taken from current production the results are just representing an one-moment-situation. Only one, the sample (K), meets the Dutch recommendations /7/ exactly.

The crushed masonry rubble from the plant (L) contains a lot of natural stone. In aggregate type (H) up to 3.5 % by weight asphalt were found. All of the samples contained very few impurities because in each case the incoming rubble was preselected on site or at the plant.

4.2 Impurities

The early German experience with recycling debris showed that preselecting and manual sorting (which in any case still cannot be avoided) leads to acceptable results if only the coarse rubble larger than 40 mm or 50 mm will be processed. Impurities seem to concentrate in the lower fractions so that they should be pre-separated. Moreover, there are some more important reasons not to reuse even the rubble fines after crushing, as for example its influence on water absorption and water demand which will be discussed later on. Only the re-use of particles larger than 2 mm can be recommended. The content of impurities must not exceed the limits for natural aggregate /3/. Additionally it is necessary to limit the content of lightweight materials like wood, plastics, paper etc. and asphalt because these may be found in rubble but some of it not in natural gravel. These contaminants which are especially typical for recycled aggregate should be limited according to /7/.

Table 1: Properties of recycled aggregates

Type		H			K		L		
Size fraction	mm	2/8	8/16	16/32	2/8	8/16	2/8	8/16	16/32
Bulk density	kg/m³	2253	2282	2256	2212	2202	2247	2416	2394
Heaped density	kg/m³	1170	1160	1140	1150	1060	1180	1260	1220
Water absorption	%	4.7	3.8	4.7	4.4	6.1	4.1	2.3	2.4

Table 2: Mix proportioning and properties of concrete

Series		H	K	L
Mix proportioning				
Cement type (blast furnace)		HOZ	HOZ	HOZ
Cement strength (28 d)	N/mm²	48	48	48
Cement content	kg/m³	320	350	320
Natural sand (0/2 mm)	kg/m³	486	559	486
Recycled natural stone (coarse)	kg/m³	294	128	925
Recycled masonry rubble	kg/m³	908	929	324
Range of gradings (DIN 1045)		A/B32	A/B16	A/B32
Total water content	kg/m³	214	226	197
Water absorption	kg/m³	54	51	37
eff. w/c-ratio		0.5	0.5	0.5
Properties of fresh concrete				
Flow	mm	370	370	350
Flow (with plasticizer)	mm	490	460	450
Compaction		1.19	1.13	1.18
Compaction (with plasticizer)		1.06	1.06	1.07
Bulk density	kg/m³	2329	2245	2323
Properties of hardened concrete (28 d)				
Bulk density (cube)	kg/m³	2210	2213	2306
Cube strength	N/mm²	45	45	47
Water penetration (DIN 1048)	mm	33	-	26
Bulk density (cylinder)	kg/m³	2274	2233	2315
Cylinder strength	N/mm²	38.8	39.8	43.4
Modulus of elasticity	N/mm²	27150	26800	28980
Bulk density (dry)	kg/m³	2103	2062	2160

4.3 Bulk density and water absorption

It was shown in /4, 5, 8/ that a useful correlation between heaped density and bulk density exists; a similar correlation is known for lightweight aggregate. New experimental data (table 1) for the pilot project in Berlin conform with the data in the literature (fig. 2). Heaped density tests are advantageous because they are very easy and rather fast whereas bulk density tests are complicated and time consuming.

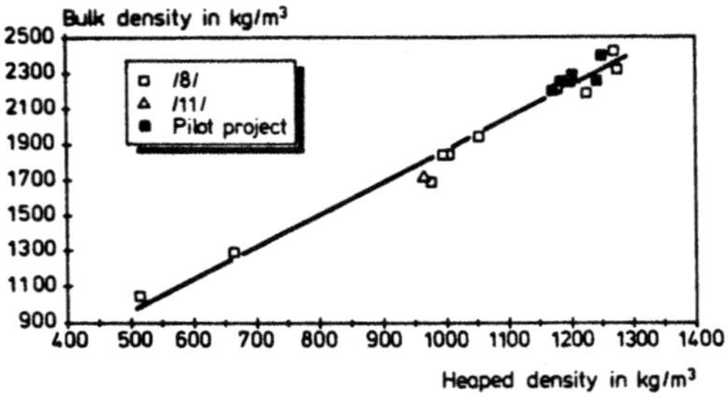

Fig. 2: Correlation between bulk density and heaped density

Heaped density tests will be suitable for quality control in order to be performed frequently. Water absorption may range from about 3 % to more than 20 % by weight depending on the kind and percentages of the constituents but also on the kind and structure of its pores. The kind and structure of the pores will especially influence the rate of absorption.

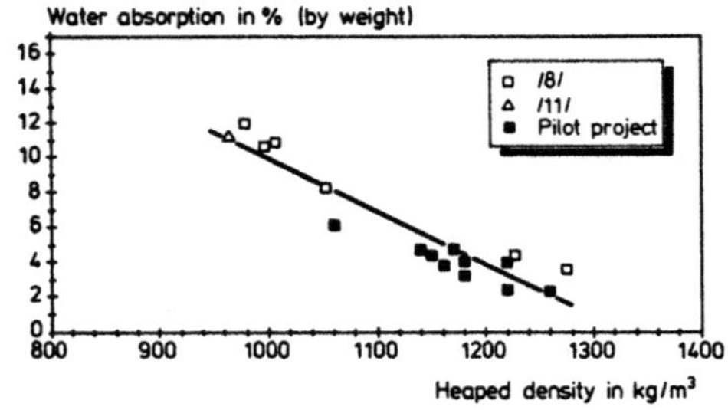

Fig. 3: Correlation between water absorption and heaped density

A higher content of crushed concrete and less crushed masonry will reduce water absorption. As water absorption is depending on bulk density and heaped density it can be estimated by quick tests as well (fig. 3 and 4).

4.4 Frost resistance

Frost resistance seems to cause severest problems if processed rubble is used. Two tested aggregates did not even pass the frost test for moderate environmental conditions according /3/. The mass loss was evidently higher when masonry rubble with very different constituents was used compared with aggregate containing more than 75 % crushed concrete. We have to consider that these short term tests may not always give real practical results but they show that the recycled material is much more sensitive than natural gravel.

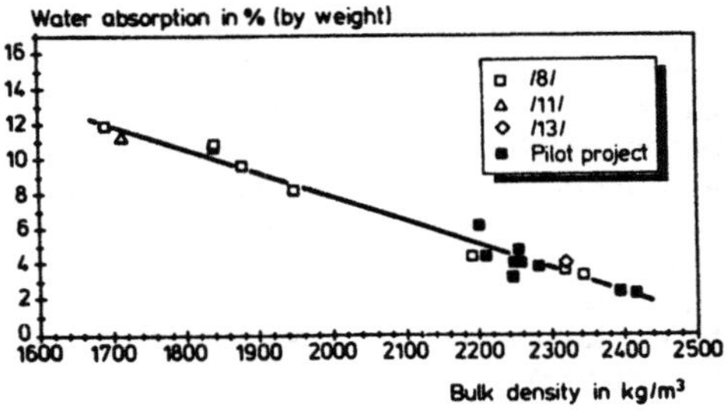

Fig. 4: Correlation between water absorption and bulk density

Crushed masonry rubble may be recommended for structural elements which are not exposed to weathering directly. The frost resistance will be improved by an increasing content of crushed concrete.

5 The properties of concrete with masonry rubble

5.1 Workability and water-cement ratio

Similar to lightweight aggregate water absorption is very significant for the mix design of concrete with recycled masonry rubble in order to adjust the water-cement ratio. The water content and the total water absorption have to be known if the aggregate is not presoaked. Evaluating the water addition on the basis of tested water absorption and water content is very difficult because the water absorption of the porous aggregate in fresh concrete differs from the absorption test especially since it is influenced by workability again. A high water content increases the rate and the amount of water absorption. So we recommend pre-wetting because this seems to be the most certain way to assure the water-cement ratio and the workability. But nevertheless, even if presoaking is used, it has to be done very careful as well. The pre-wetting procedure must take enough time and guarantee a uniform distribution of water.

The literature from the times after the Second World War (see reviews /4, 5/) shows that the total water demand was more often estimated than tested. An estimate of 22 % up to 30 % by weight related to the dry aggregate was used. No differences were made between the water-absorptive capacity and water demand which is influenced by the size distribution and shape of the particles. The water addition was judged by workability tests. It was checked whether workability had changed within half an hour after mixing or not. Some authors proposed that the aggregate should be pre-wetted with a certain amount of water during half an hour. Afterwards as much water was to be added as was needed to achieve a certain workability. Even these procedures could not guarantee the required or calculated water-cement ratio. Very often the effective ratio was higher or even lower.

Plasticizing admixtures on melamine basis can be used to improve workability or to reduce water addition. But our tests show that the effect of improvement lasts only for a rather short time (not even 30 minutes). This may depend on the type of plasticizing admixture but probably it depends on the above mentioned mutual relation between workability and water absorption, too, though pre-wetted (but probably not completely presoaked) aggregates were used. Plasticizing admixtures reducing the surface tension of water (wetting agents) are unsuitable because they increase water absorption /9/.Unexpectedly five minutes of premixing the aggregate with a laboratory force mixer did not impair the workablitity of fresh concrete. On the opposite, workability seemed to be slightly improved. This result may be explained by moderate abrasion and refining with the grading curve moving upwards only slightly. It still remains within the range of good grading but probably the particles become better shaped by abrasion.

5.2 Strength
Concrete with recycled aggregate has a lower compressive strength than concrete with natural aggregates. It is already known that the difference amounts to about 10 to 20 % if crushed concrete rubble is used /4, 5, 6, 7, 10, 13/. An increasing percentage of other weaker particles such as some porous types of brick may reduce strength. In spite of this the design of concrete compressive strengths up to 30 N/mm² (B 25) will cause no problems so that the current mix design for normal concrete or lightweight concrete can be used. Difficulties may occur if higher strength classes are required.

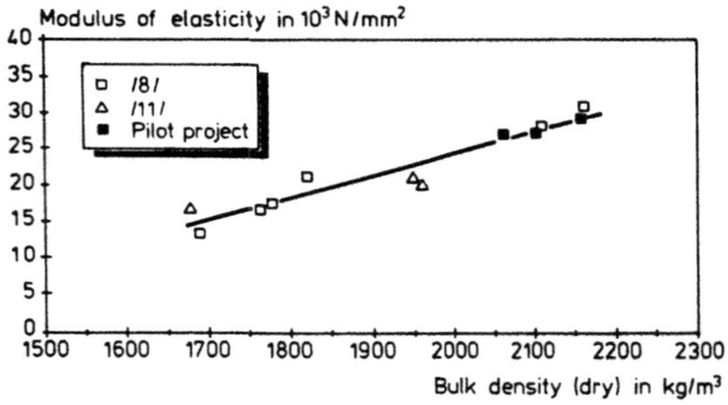

Fig. 5: Correlation between modulus of elasticity and bulk density

5.3 Modulus of elasticity
The deformation behavior differs from that of normal concrete /4, 5, 7, 10, 11/. Variations caused by the aggregate can be equalized by mix design only within a small range. Nevertheless, the modulus of elasticity is predictable as there exists a rather strong correlation between concrete bulk density and the modulus of elasticity (fig. 5) but also between strength and the modulus of elasticity.

The modulus of elasticity for concretes with the same cement paste (table 2) shows that the influence of the type of the aggregate constituents is significant. A higher percentage of natural gravel reduces the deformability.

5.4 Durability
According to German Standards a sufficient frost resistance of concrete will be obtained if the water permeability test shows a penetration depth less than 50 mm and if the water-cement ratio is lower than 0.60. The aggregates have to be frost resistant too. What could be done if

the aggregates are not frost resistant? Since the frost resistance is a matter of porosity, would it help if the porosity of cement paste is reduced by a water cement-ratio of 0.50? Our tests showed that in fact the average water penetration was between 21 and 33 mm (table 1), but one specimen of one serie showed a maximum penetration depth of 55 mm locally. There was a large sized aggregate particle (mortar) with a large quantity of pores touching the surface. So we must learn that a better cement paste quality may improve the density and probably the frost resistance but it cannot prevent local defects. The authors of /6/ confirm this by frost-thawing tests. Additionally they found out that the frost-resistance of concrete with masonry rubble is much more deviating than that of normal concrete. As carbonation is influenced by similar factors the covering of the reinforcement has to be chosen more carefully than for normal concrete. But this is true for lightweight concrete with porous aggregates at all. After /12/ a sufficient frost-thawing and deicing resistance may be provided if the crushed concrete and the new paste as well are air entrained.

Such old concrete which is air entrained will be found for example in road pavements. Though concrete from pavements may recommend itself to be recycled because of its rather high and constant quality as well as the low degree of reinforcing, at least one disadvantage must be taken into account. In Germany, deicing salts have been used quite extensively. Chlorides penetrated into the concrete and may exceed the limit of 0.4 % (German standard value for portland cement) related to the cement mass. Chloride contents above this limit are regarded as harmful for the reinforcement. But if only small surface layers contain excessive amounts of chloride it should be discussed whether corrosion can be avoided by crushing and equalizing.

6 Conclusions

If recycling building materials is regarded as a closed loop system all valuable constituents in the waste may not be dissipated by applications of secondary importance (e.g. land fill). As far as it can be afforded from an economical point of view, processed rubble should be employed on the highest possible level of utilisation. The re-use of building waste as aggregate for concrete seems to be especially prosperous and may be regarded as the highest level of refining. Recycled aggregate will be sufficient if only preselected or cleaned material with particle size above 2 mm is re-used and if the percentages of the impurities conform to the standards for normal or lightweight aggregate and/or and the Dutch recommendations for recycled aggregate /7/. Combinations of natural aggregate and recycled aggregate may be considered. Recycled aggregate concrete can be used for applications with reduced requirements especially with regard to frost resistance. Valuable preselected material like concrete from pavements can be re-used for advanced purposes if the amount of impurities like chlorides is tolerable. Future experimental research should concentrate on homogenizing the processed rubble and on presoaking procedures in order to reduce possible property deviations. The influence of recycled aggregate on mechanical properties is much better explored than its influence on durability.

References

/1/ Lamprecht, H.-O.: OPUS CAEMENTITIUM - Bautechnik der Römer. Beton-Verlag, 1984 (in German).

/2/ DIN 4163, Ausgabe 02.1951: Ziegelsplittbeton. Bestimmungen für die Herstellung und Verwendung (in German, withdrawn Standard).

/3/ DIN 4226, Ausgabe 04.1983: Zuschlag für Beton (in German).

/4/ Schulz, R.-R., Wesche, K.: Recycling von Baurestmassen - Ein Beitrag zur Kostendämpfung im Bauwesen. Forschungsbericht, gefördert durch das Bundesministerium für Raumordnung, Bauwesen und Städtebau. RWTH Aachen, Aug. 1986, IRB-Verlag Stuttgart (in German).

/5/ Mulheron, M; Hendriks, C.F.; Schulz, R.-R.: Recycled Masonry Rubble as an Aggregate for Concrete. State-of-the-art report, developments 1945 -1985. RILEM TC-37 DRC (Demolition and Recycling of Concrete), Second Draft, Feb. 1986.

/6/ Ivanyi, G.; Lardi, R.; Eßer, A.: Recycling Beton, Zuschlag aus aufbereitetem Bauschutt. Forschungsberichte aus dem Fachbereich Bauwesen, Heft 33, Universität-Gesamthochschule-Essen, Sept. 1985 (in German).

/7/ Stichting voor onderzoek, voorschriften en kwaliteitseisen op het gebied van beton (CUR-VB): Aanbeveling 4, Betonpuingranulaat als toeslagmaterial voor beton. Aanbeveling 5: Metselwerkpuingranulaat als toeslagmaterial voor beton. Zoetermeer, november 1984. Aus: CUR-rapport 125 (s. /8/, in Dutch).

/8/ CUR-rapport 125: betonpuingranulaat en metselwerkpuingranulaat als toeslagmateriaal voor beton. Civieltechnisch centrum uitvoering research en regelgeving (CUR), NL, Gouda, 1986 (in Dutch).

/9/ Sommer, H.: Recycling von Beton - Wiederverwertung im Deckenbau. In: Straße und Autobahn 35 (1984), No. 5, pp. 187 - 189 (in German).

/10/ Hansen, T.C.: Recycled aggregates and recycled aggregate concrete, second state-of-the-art report, developments 1978 - 1985. RILEM TC-37 DRC (Demolition and Recycling of Concrete) Materials and Structures 19 (1986), No. 111, pp. 201 - 246.

/11/ Akhtaruzzaman, A. A.; Hasnat, A.: Properties of Concrete Using Crushed Brick as Aggregate. In: Concrete International (1983), Feb., pp. 58 - 63

/12/ Kleiser, K.: Wiederverwendung von Betonschutt als Betonzuschlag. Vortrag im Rahmen der Fachveranstaltung: Aufbereitung und Wiederverwendung von Bauschutt. Haus der Technik, Essen 4. Dez. 1986 (in German).

/13/ Wesche, K.; Schulz, R.-R.: Beton aus aufbereitetem Albeton - Technologie und Eigenschaften. In: beton 32 (1982), No. 2, pp. 64 - 68, No. 3, pp. 108 - 112 (in German).

THE RECYCLING OF DEMOLITION DEBRIS: CURRENT PRACTICE, PRODUCTS AND STANDARDS IN THE UNITED KINGDOM

M MULHERON
Department of Civil Engineering, University of Surrey

Abstract
This paper presents a review of current practice and products concerning the recycling of demolition debris in the United Kingdom. The types of recycling plants currently in operation are outlined and an indication is given of the types of recycled aggregate products being produced. Current and potential uses for such products are discussed in relation to current British Standards and Specifications covering materials for new construction. Cost comparisons between recycled aggregates and the natural aggregates with which they compete are also presented.
Key words: Standards, Materials Specifications, Recycling plants, Aggregate properties, Recycled products, Economics of recycling, Uses of recycled aggregates.

1. Introduction

In recent years there has been a growth of awareness in the United Kingdom (U.K.) for the need to conserve natural resources and recycle or reclaim those materials which are in short supply. Indeed according to a recent publication of the Trade and Industry Committee of the House of Commons (1984); "the benefits of recycling are obvious: rarely do environmental and economic factors so unambiguously support the same goal".

Current estimates, Environmental Resources Ltd (1980), indicate that the amount of demolition debris dumped at land-fill sites in the U.K. is in excess of 20 million tonnes per annum. The bulk of this material is concrete (50-55%) and masonry (30-40%) with only small percentages of other materials such as metals, glass and timber. Lindsell and Mulheron (1985) have calculated that if some of this debris was recycled as useful aggregate then it could reduce the demand on natural resources by as much as 10%. This would help supplement the supply of natural aggregates, extend the life of existing quarries, and at the same time reduce the rate of consumption of space in land-fill sites.

In 1985 the Institute of Demolition Engineers commissioned a study of demolition debris and its potential for recycling and reuse, Lindsell and Mulheron (1985). This revealed that many demolition contractors are already involved in recycling demolition debris in some way. In many cases this simply involves the use of mixed debris for raising the level of a site or the provision of temporary access roads. Much of the material put to such uses goes unrecorded, making it likely that the 20 million tonnes of demolition debris thought to be available for recycling each year is a conservative estimate.

For those demolition contractors involved in the recycling of demolition debris it is essentially a profit motivated operation resulting from the high cost of transporting and tipping demolition debris in urban areas. Thus, where the value of the recycled material (minus the cost of reprocessing) returns a profit over the cost of simply tipping the debris then recycling becomes attractive. Obviously any recycling operation must be feasible given the confines of the site and will only be carried out where it does not interfere with the main time schedule of the demolition process.

The main areas currently supporting recycling in the U.K. are London and the South-east with practically no large scale recycling being carried out in Wales or Scotland. In part this reflects the fact that over 70% of the total quantity of demolition debris in the U.K. is produced in England but also reflects the fact that recycling is limited to those areas with few natural aggregates.

Typical costs for the dumping of demolition debris in areas supporting recycling are shown in Table 1. which shows that they can vary by a factor of 6 even within small areas. This is because sites for the disposal of domestic refuse charge high prices for accepting debris as it takes up space that could otherwise be used for domestic waste. In contrast land-fill sites for later land reuse are limited by their licence to only accept inert materials. As a result such sites charge relatively low prices for clean debris but will often not accept debris heavily contaminated by timber.

Table 1. Tipping costs for general demolition debris, 1985 prices.

Location	Typical charge per loose cubic metre (£)
Chippenham, Wilts.	1.00
Docklands, London.	No tipping allowed
Erith, Kent.	1.30
Halesowen, West Midlands.	2.60
Leicester, Leicestershire.	0.50
Portsmouth, Hants.	0.60 - 3.50
Salford, Lancashire.	2.50

Whilst high tipping costs encourage the recycling of demolition debris it is also necessary that there is a market for the recycled product. The main markets for recycled aggregates are as material for fill, earthworks and sub-base layers in construction. For such applications recycled aggregates must compete with natural aggregates the prices of which depend on the cost of transport and are at their highest in large conurbations far from a source of natural aggregate.

Typical prices for recycled aggregates produced by plants in England are shown in Table 2. Crushed concrete produced for use as a granular sub-base material is sold at £2.00-£5.00 per tonne. This may be compared with prices for natural aggregates used for similar applications shown in Table 3 which indicates that a price differential of between £1.00 and £4.00 is usual. Lindsell and Mulheron (1985) found that for recycled aggregates to compete with natural aggregates a minimum price differential of between £1.00 and £2.00 per tonne is required.

Table 2. Costs of recycled materials from demolition debris, 1985 prices.

LOCATION	MATERIAL	COST PER TONNE (£)
Chippenham, Wilts.	Concrete & Brick	2.50
	Brick	2.00
Docklands, London.	Concrete & Brick	3.00
Erith, Kent.	Concrete & Brick	4.00
Halesowen, West Midlands.	Concrete & Brick	4.50
Leicester, Leicestershire.	Concrete	2.00
Portsmouth, Hants.	Concrete & Brick	2.50 - 4.75
Salford, Lancashire.	Concrete	3.50 - 4.00
	Brick	2.00 - 3.00

Table 3. Costs of natural aggregates in areas supporting recycling, 1985 prices.

LOCATION	MATERIAL	COST PER TONNE (£)
Chippenham, Wilts.	Limestone	3.60
Docklands, London.	Gravel	6.50 - 9.00
	Limestone	7.00 - 8.00
Erith, Kent.	Natural stone	7.00
Halesowen, West Midlands.	Natural stone	4.00 - 5.50
Leicester, Leicestershire.	Natural stone	4.20
Portsmouth, Hants.	Gravel	2.70 - 4.50
	Limestone	6.00 - 8.00
Salford, Lancashire.	Natural stone	5.00
	Limestone	6.50

Whatever the economics of the recycling operation, unless the recycled product is acceptable to local authorities its use will be severely limited. Research results, Hansen (1987) and Hendriks and Schultz (1988), and past experience, Newman (1946), all indicate that recycled brick and concrete aggregate can be used successfully for a wide range of applications. Indeed in many of the large conurbations in the U.K., such as London and Manchester, crushed concrete and brick are accepted for use as fill and sub-base materials.

2. Production of recycled aggregates

In reviewing the recycling of brick and concrete demolition debris, three separate types of debris may be identified, clean brick or masonry, clean concrete - plain or reinforced, and mixed demolition rubble. By far the most common of these in the U.K. is mixed rubble, since the time constraints of a demolition contract rarely allow the contractor to selectively demolish and recover clean material.

Plants for the production of recycled aggregates can incorporate a variety of crushers and sorting devices. When processing clean brick or concrete debris a simple crushing and screening plant is capable of producing an acceptable product. The processing of mixed and contaminated debris requires additional sorting techniques. Low levels of contaminants can be dealt with by hand separation methods whilst in more sophisticated recycling, lightweight impurities are removed by water-floatation or air-sieving. Gross contamination by wood or plastic is undesirable and in many cases heavily contaminated debris is not recycled.

In general, recycling plants in the U.K. are restricted to single crusher 'on-site' installations and usually consist of a single jaw crusher working with associated sieves, and sorting devices. Such plants are capable of producing a crushed and graded material but only used where there is sufficient debris on the site to justify the expense of setting up the plant. The quality of the product depends on the type of debris being crushed since reprocessing is limited to hand sorting and electromagnetic separation of iron and steel. The crushed debris is often used as fill material on the same site.

The only true 'fixed-site' crusher plants currently operating in the U.K. are located in London and Manchester. These plants typically combine two jaw crushers, one primary and one secondary, and are capable of producing a range of graded products. The separation techniques used are normally limited to self-cleaning electromagnets, sieving and hand-sorting, although recently a number of experiments have been made with simple float-sink tanks to help remove timber fragments and excess fines. Such installations have the advantage of being able to accept a more mixed and contaminated demolition debris and yet still produce a range of clean, graded aggregates.

Almost all of the recycling plants currently operating in the U.K. utilize jaw crushers whereas recycling plants in the rest of Europe favour the use of impact crushers. The fundamental difference between the jaw and impact crusher lies in the method by which the material is crushed. According to Boesman (1985) this produces a marked effect on the particle shape and size distribution of the crushed product. Normally impact crushers produce a more angular product than jaw crushers and also have a larger reduction factor. This is important because for the same maximum size of coarse recycled aggregate an impact crusher will generate twice the amount of fines produced by a jaw crusher. Apart from the different particle shape and size distribution, the physical properties of recycled aggregates such as specific gravity, water absorption and abrasion loss percentage are not significantly affected by the type of crusher used.

The majority of crushers and other equipment currently in use in the U.K. for the production of recycled aggregates has been obtained second-hand from quarry plant manufacturers and modified to meet the requirements of demolition debris. In part this helps explain why the quality of recycled aggregates produced in the U.K. is generally lower than that produced in some parts of Europe.

3. Recycled aggregate products

Depending on the type of debris being processed, and the type of recycling plant available, a range of recycled aggregates can be produced. For comparison purposes it is possible to classify this range of products into 4 main categories:-

(a) Crushed demolition debris - mixed crushed concrete and brick that has been screened and sorted to remove excessive contamination.
(b) Clean graded mixed debris - crushed and graded concrete and brick with little or no contamination.
(c) Clean graded brick - crushed and graded brick containing less than 5% of concrete or stony material and little or no contaminants.
(d) Clean graded concrete - crushed and graded concrete containing less than 5% brick or stony material and little or no contaminants.

By far the largest quantity of recycled aggregates produced in the U.K. falls into the first category. The major differences between recycled and natural aggregates can be summarized as follows;

(a) Particle shape and surface texture. - Recycled aggregates tend to have a particle shape which is more irregular than natural aggregates and also possess a coarser surface texture.
(b) Density. - The density of recycled aggregates is usually lower than that of natural aggregates due to the presence of old mortar, bricks and other low density material.
(c) Water absorption. - The most marked difference in physical properties of recycled aggregates when compared to conventional aggregates is their higher water absorption. Mulheron and O'Mahony (1987) have found water absorptions for coarse recycled concrete aggregates ranging from 5.3% to 8.3% compared 1.5% to 3.1% for equivalent natural gravel aggregates. These values are in good agreement with those obtained by Hansen and Narud (1983).
(d) Durability. - According to Hansen (1986) there are conflicting reports concerning the durability of recycled aggregates. Tests on recycled aggregates in the U.K., Mulheron (1986) and Mulheron and O'Mahony (1987), suggest that when subjected to alternating freeze-thaw conditions unbound recycled aggregates are less durable than natural river gravels. Interestingly when concretes made from these aggregates were tested under the same conditions of freezing and thawing it was observed that the recycled aggregate concrete exhibited the superior durability.

4. Uses of recycled aggregates

Simply producing a clean, crushed and graded material, is not sufficient to ensure effective recycling. The recycled material produced must be suitable for specific applications and thus should comply with certain grading limits, contain minimal levels of contaminants and meet other requirements of stability and durability. Once the rubble has been crushed, sieved and decontaminated it can find application as; i). general bulk fill, ii). fill in drainage projects, iii). Sub-base material in road construction, and iv). aggregate for new concrete manufacture. The suitability of recycled aggregate for such uses in shown in Table 4. It should be noted that whilst the technical acceptance of recycled concrete and brick aggregates depends on the results of laboratory tests and field trials, a major barrier to preventing their more widespread acceptance is the lack of Standards covering these materials.

There are currently no Standards in the U.K. specifically covering the use of recycled aggregates and so recycled products can only be compared with existing Specifications developed for natural aggregates. Such a comparison can be totally inappropriate and tends to inhibit the future development of recycled materials. However, according to one recently introduced British Standard Guide, BS 6543 (1985), both brick and concrete aggregates can be suitable for a range of applications and even suggests that they can be used as aggregates for the production of new concrete.

4.1 Current uses in the United Kingdom

4.1.1 Fill material

Foundations - There are currently no British Standards for materials used as fill in building foundations although the choice and use of materials has been outlined by the Building Research Establishment

Table 4. Suitability of recycled aggregates for various applications.

Recycled Aggregate Category	General bulk Fill	Fill in Drainage Projects	Material for road construction	New concrete manufacture
Crushed Demolition Debris.	Suitable	Usually suitable	Not usually suitable	Not suitable
Graded Mixed Debris.	Suitable	Usually suitable	Suitable in some cases	Suitable in some cases
Clean, Graded Brick.*	Highly suitable	Suitable	Usually suitable	Suitable in some cases
Clean Graded Concrete.	Highly suitable	Highly suitable	Suitable	Usually suitable

* Suitability depends on type of brick

(1983). Ideally a material for use as fill should be a hard granular material with a fairly large particle size that consolidates easily and remains free draining. It should also be chemically inert and not subject to significant changes in dimension with changing moisture content.

Clean graded concrete meets these requirements admirably and is used in many countries, including the U.K., in the construction of foundations for houses, garages and other light buildings. Clean graded brick is also used as a fill material in such applications provided it is of sufficient hardness and durability. Whilst the use of these materials has been quite widely accepted in the U.K. in the past, a number of local authorities have recently specified that material for foundations must be a clean quarry stone. As a direct result of this change in attitude demolition contractors in some areas are being forced to dump brick based debris even though it has been used successfully for many years.

General bulk fill - Graded, mixed debris and crushed demolition debris are usually suitable as general fill but debris with a high sulphate or timber content should be avoided for critical applications. Much of the inferior quality material is used for less demanding applications such as landscaping, levelling and the construction of acoustic barriers.

4.1.2 Material for road construction

The requirements for material used in road construction in the U.K. are set out in the 'Specification for Highway Works', Department of Transport (1986). At present the only recycled material specifically included in this specification is crushed concrete which is allowed as granular fill for a wide range of applications such as;

- Drainage works, for use as permeable backing to earth retaining structures, material for filter drains, and backfill to pipe bays and above pipe surround material.
- Earthworks, such as fill to structures, drainage layers to reinforced earth structures, bedding material for buried steel structures, and as unbound or cement/lime bound capping layers.
- Road base and sub-base layers, as Type 1 or 2 granular sub-base material.

Despite the appearance of crushed concrete in the Specification as being acceptable as a selected granular fill for such applications its use often meets resistance. This results from the perception of many people in the U.K. that recycled aggregates are of low quality and variable performance.

Whilst neither recycled brick or mixed debris are mentioned in the current Specification it does not actually exclude their use. Indeed an analysis of the current requirements of materials for earthworks suggests that both materials are capable of meeting the requirements for general granular fill, classes 1(A,B,C), Landscape fill, class 4, selected granular fill, classes 6(E,F), and stabilized materials, class 9(A).

4.2 Potential uses

4.2.1 Aggregate for cement bound material

Research results and practical experiance have shown that given the correct grading and other properties both crushed concrete and brick can be successfully used for the production of both cement bound materials and lean concrete. The current Specification for cement bound materials includes 4 categories of material with minimum strengths at 7 days ranging from 4 MPa. to 15 MPa. Of these materials only CBM1 and CBM2 can be manufactured with recycled aggregates since the Specification for CBM3, CBM4, and all categories of wet lean concrete requires the aggregate to be either a naturally occurring material complying with BS 882 (1983) or crushed air-cooled blastfurnace slag complying with BS 1047 (1983). This represents a change from previous editions of the Specification which required aggregates for cement bound materials to be a 'washed or processed granular material' which included the using of recycled materials.

4.2.2 Aggregate for the production of new concrete

To be suitable for the production of new concrete an aggregate must meet a number of requirements. Firstly it must be sufficiently strong for the grade of concrete required and possess good dimensional stability. Secondly, it must not react with either the cement or reinforcing steel nor should it contain any potentially reactive impuirites. Finally the aggregate should have a suitable particle shape and grading to produce a mix with acceptable workability.

The production and properties of concrete made with recycled concrete and brick aggregate has been the subject of a considerable research effort much of which has been reviewed by Hansen (1986) and Hendriks and Schulz (1988). Based on the results of laboratory investigations and field trials it has been found that clean brick and concrete aggregates can produce a concrete with acceptable workability and strength.

A major barrier to the accceptance of recycled aggregates for the manufacture of concrete is the difficulty in producing a material that conforms with existing Standards. At present neither crushed concrete of brick comply with existing British Standards covering either aggregates from natural sources for concrete, BS 882 (1983), or lightweight aggregates for concrete, BS 3797 (1982). Indeed whilst British Standards have been produced to cover the use of manufactured aggregates from a number of sources, such as air-cooled blastfurnace slag aggregate, BS 1047 (1983), foamed and expanded blastfurnace slag lightweight aggregate, BS 877 (1977), it is open to question whether a Standard will be introduced to cover the use of recycled concrete or brick aggregates.

5. Conclusions

At present, the recycling of demolition debris is carried out by a limited number of demolition contractors and only when the economics of the operation are favourable. In almost all cases the level of processing that occurs is limited to simple crushing, followed by sieving and hand sorting. As a result the quality of the recycled product is largely determined by the type of debris being processed and is normally only suitable for fill and sub-base appplications. There are currently three major barriers to improving the quality of the recycled materials currently being produced in the U.K.;

(a) Lack of incentive - since there are no Standards covering the production and use of recycled aggregates contractors in the U.K. have little incentive to improve the quality of the materials that they produce. Similarly since recycled aggregates are not included in British Standards for either normal or lean mix concrete, there is no incentive to produce a recycled aggregate suitable for such requirements.

(b) Cost - most contractors are only able to afford the use of a single crusher working with associated sieves and electromagnets and such an arrangement is limited in the quality of material it can produce.

(c) Attitudes to recycled aggregate products - many people associate recycled materials from demolition debris with a low quality and variable performance. In practice this is often untrue since experience has shown that suitable processing recycled aggregates can be used for a wide range of applications. Of course having produced a recycled aggregate it is important to ensure that its use is based on a sound understanding of its properties. For example, whilst crushed concrete can be used as an aggregate for the production of new concrete, it should not be used as a straight substitute for natural aggregates. Rather, if optimum workability and strength are to be obtained, a mix must be designed based on the specific properties of the recycled aggregate. It is this recognition of recycled aggregates as materials with their own distinct properties which has led a number of countries to propose separate Standards for their use.

It must also be remembered that people involved in specifying materials for construction projects prefer to use materials with a proven performance and as a result many architects and consultants in the U.K. will not specify recycled materials. This understandable reluctance to use what are percieved as low quality alternatives is often the result of a lack of well documented laboratory tests and field trials. It appears likely that this situation will continue until either new Standards are produced for recycled materials, or current Standards and Specifications are amended to include recycled materials.

As a result of the current lack of Standards for recycled materials the Institute of Demolition Engineers are currently sponsoring fundamental research on recycled materials. The basic mechanical, physical, and chemical properties of the aggregates are being assessed, and the materials subjected to a number of full-scale trials. Although results from other countries provide a useful database for such tests it is felt important to validate any conclusions using materials produced and tested under British conditions if they are to be used in developing future Specifications and Standards for recycled materials in the U.K.

Acknowledgements
This paper is based on the report 'Recycling of Demolition Debris and is published with the kind permission of the Institute of Demolition Engineers.

References

Boesman, B., (1985) Crushing and separating techniques for demolition material. EDA/RILEM Conference on Demo-Recycling, Part II - Reuse of concrete and brick materials, Rotterdam
Building Research Establishment (1983) Hardcore, Digest 276, Department of the Environment.
BS 877 (1977) Part 2: Foamed or expanded blastfurnace slag light-weight aggregate for concrete. British Standards Institution.
BS 882 (1983) Specification for aggregates from natural sources for concrete. British Standards Institution.
BS 1047 (1983) Specification for air-cooled blastfurnace slag aggregate for use in construction. British Standards Institution.
BS 3797 (1982) Part 2: Specification for lightweight aggregates for concrete. British Standards Institution.
BS 6543 (1985) Guide to use of industrial by-products and waste materials in building and civil engineering. British Standards Institution.
Department of Transport (1986) Specification for Highway Works. Her Majesty's Stationary Office.
Environmental Resources Ltd. (1980) Demolition waste. Construction Press Ltd., Lancaster, England. (ISBN 0-86095-865-5)
Hansen, T.C. (1986) Second state-of-the-art report on recycled aggregates and recycled aggregate concrete. Materials and Structures, 9 (73), p5.
Hansen, T.C. and Narud, H. (1983) Strength of recycled concrete made from crushed concrete coarse aggregate. Concrete International - Design and Construction, 5(1).
Hendriks, C.F. and Schulz, R.R. (1988) State-of-the-art report on recycled brick rubble as an aggregate for the production of new concrete. Submitted for publication in Materials and Structures.
House of Commons (1984) The wealth of waste - fourth report (session 1983-84), Trade and Industry Committee of the House of Commons, London, HMSO.
Lindsell, P and Mulheron, M. (1985) Recycling of demolition debris. The Institue of Demolition Engineers.
Mulheron, M. (1986) A preliminary study of recycled aggregates. The Institute of Demolition Engineers.
Mulheron, M. and O'Mahoney, M. (1987) Recycled aggregates: Properties and performance. The Institute of Demolition Engineers.
Newman, A.J. (1946) The utilization of brick rubble from demolished shelters as aggregates for concrete. Institute of Municipal Engineers Jounal, 73(2) pp 113-121.

RECYCLING OF CONCRETE AND DEMOLITION WASTE IN THE U.K.

A. TREVORROW, DR. H. JOYNES and DR. P. J. WAINWRIGHT.
Department of Building & Environmental Health, Trent Polytechnic, Nottingham.

Abstract
This paper highlights how the UK has managed the introduction of the recycling of demolition waste into the economy in the last five or six years compared with the philosophies and evolved policies of other European countries, in particular The Netherlands.

To measure the extent of the UK's progress the background and structure of the industry will be discussed. The type and quality of the products, both specified and achieved, will be examined in relation to their use in the road construction industry. The means of producing these recycled products will be investigated and two case studies given to highlight some of the difficulties being met by both clients and the demolition industry in recycling demolition waste.

Finally, a summary giving the points which need further discussion so that the recycling of demolition waste will benefit the construction industry and provide a further cost effective national resource.

1. Introduction

It was the Dutch, Belgians and West Germans in 1976 who first encouraged the EEC to commission a report on the

> "examination of the arisings end uses and disposal of demolition waste in Europe and the potential for further recovery of materials from these wastes." (1)

The report predicted, that the quantity of demolition material would in fact double by the year 2000 and triple by the year 2020 (1). Further, it suggested that there would be a reversal of the proportions of the main constituents, bricks and concrete, from a concrete content of 35% presently to approximately 80% by the year 2000. For this reason and to alleviate the acute shortage of raw materials the Dutch government proposed a whole new policy of recycling encompassing not only demolition waste but all waste materials.

2. Background from the UK

At the same time the UK was not deficient in raw materials as it had ample supplies of a wide variety of mineral deposits from which, under licence, raw materials for concrete and the construction industry could be found. Most of these natural resources provide high quality materials at very competitive prices of £2-£6 per tonne. (3) Therefore, the interest in the EEC report was minimal, being restricted to a few design engineers, demolition contractors and research establishments.

What has arisen since, is that the reports prediction of the increase in concrete demolition waste has already begun to show itself here in the UK. The increase has come about because in the 1960's and 70's the UK went through the re-housing programme led by Local Authorities and the architectural philosophy of the time to deposit people in tall multi-storey buildings constructed of concrete frames and precast concrete panels, rather than the traditional brick built homes.

These concrete structures, due to expediency in many cases, were not built with the same quality assurance compared with the standards of structures today. The Local Authorities who commissioned these new homes were faced almost from the start with maintenance and social problems, and these problems have led to the decisions in the late 1980's to demolishing these blocks after only 15-20 years of their expected design life. Further to this, the main industries of many areas have fallen into economic decline and the buildings have been demolished.

Finally the UK road system, constructed of many miles of concrete, has had to be replaced due to the increase in traffic, caused by the demand from the market place on the transport industry. Therefore the UK finds itself with certain problems:

(1) Reinforced concrete demolition is increasing faster than thought 10 years ago.

(2) Already in certain areas of Britain tips for disposal of waste materials are very expensive, inaccessible or non-existent. (2)

(3) In certain areas some aggregates cannot be obtained as easily as they were 10-15 years ago. (2)

Recycling can be used as part of the solution to some of these problems. But one must consider which techniques and standards should be adopted by the recycling industry so that it may become competitive and useful to the nation's construction industry.

These problems are exacerbated by the reluctance of the industry to refine their products in terms of constituents, as the cost is not justified by way of return on the investment required. Consequently the UK produces a large volume of low grade mixed demolition fill material.

3. Specification

The standards that are being used in the UK are basically those established for road foundations, namely D.O.T. type 1 and 2 (5) which allow crushed concrete to be used. Therefore specifiers and contractors have successfully used a recycled material. Conversely ad hoc specifications are written by engineers to suit individual contracts when a recycled material becomes available, eg as a formation level capping course. (3)

4. The structure of the demolition recycling section of industry

The recycling industry has evolved from roots in quarrying and aggregate extractions, of which the UK can be proud, and now consists of four broad areas of operation.

(1) Demolition contractors who employ crushers to control the costs of their demolition work in terms of transportation efficiency. The Institute of Demolition Engineers is encouraging an interest in developing a specification for recycled mixed demolition waste. The cost of producing recycled material in the area is approximately £1.75-£3.00 per tonne.

(2) Crushing sub-contractors are hired by both demolition contractors and national aggregate suppliers, with a wide variety of jaw, roller, impact and core crushers. This enables specific standards to be met, (4, 6, 7) with a cost of production to the client of approximately £2.00-£3.50 per tonne.

(3) Aggregate suppliers will also utilize demolition waste in areas where their natural material is scarce. In this instance the cost of production is approximately £1.75-£2.80 per tonne.

(4) A number of building contractors undertake operating hired crushers to utilize the demolition waste on their sites and then use the good draining qualities of the recycled material for the temporary roads, capping courses or as an infill material (2, 3, 4). The cost of production in this situation is approximately £2.50-£4.00 per tonne. All these sections of the industry are managing a new product and are learning fast with the help of architects, engineers, the Health and Safety executives and from the experience of our European partners.

5. Site production

The case studies have been chosen & represent the more common types of problems found with the demolition and recycling of waste on sites in the UK.

A typical site set up to produce a crusher run material in the UK consists of the following items of plant:

(i) 360° tracked hydraulic backactor
(ii) jaw crusher, single or double toggle
(iii) straight or swing conveyor with screen
(iv) tracked or rubber wheeled loader.

This plant would be suitable for contracts ranging from 40,000-150,000 tonnes, where the demolition waste is delivered to site in small (i.e. 0.250T) sections.

6. Case Study No 1 (Plate 1)

This contract consisted of the demolition of 700 dwellings, within the city of Nottingham. The structure was of in-situ and precast concrete panels. The process of recycling was restricted by safety and noise restraints imposed by the city authority. For the means of crushing the concrete panels etc where transported to a tip some 8 miles away. Extra transport costs had to be accounted for and further site breaking was incorporated to enhance the load capacity of the lorries.

The crushing facilities at the tip were a straightforward crusher run set up, which was able to attain a ready saleable D.O.T. type 2 standard recycled aggregate, at a cost of approximately £2.50/tonne and a resale value of £3.15/tonne. (4)

7. Case Study No 2

This contract consisted of the demolition of 1200 dwellings, again within the city of Nottingham. The structures were of low and multi storey precast concrete panel frames. Crushing on site or at a tip was refused because of nuisance related reasons by the local authority. This resulted in a total cost to the project of tipping the material and there was no credit from the value of the potential recycled material.

8. Observations

The situation in the UK may be summarised as follows:

(a) there is some confliction between the natural aggregate suppliers and the demolition recycling industry.

(b) unless demolition contractors are allowed to crush on sites or a regulated location, the disposal of the increasing tonnages of both plain and reinforced concrete will become more acute.

(c) guide lines relating to the recycling of demolition waste are clearly inconsistent.

(d) competition appears to be inhibiting research into refining recycling production. Although the Institution of Demolition Engineers is striving to achieve a draft specification. (8)

(e) the plant used has developed little from its origins. Site production is based on the mobile plant available.

(f) the most popular types of plant used are jaw crushers: cone crushers and or another jaw crusher is provided where the quality of the product is important.

(g) fixed installations seem to be three to five years away and these would be in large conurbations (over 1,000,000 population) as predicted in the EEC report (1). Private finance may be required for fixed installations, or they could be incorporated in with schemes similar to London waste regulation Authority (9) which at present moves domestic and hazardous waste out of London.

(h) finance for the fixed installations could be difficult to attract due to the high risk factor on a project where there is a low priced product marketed against a well established, very competitive, high quality natural aggregate.

9. Conclusion

The demolition and recycling section of the construction industry can clearly provide the UK with a valuable resource. This has been proven in the Netherlands, as in other European countries. The lack of organisation and direction in developing this area present the whole recycling sector with a mass of problems which it cannot hope to solve on its own.

The policy statement by the Dutch government set their recycling industry off on its challenging way with a nationally backed incentive. This appears to be what the UK lacks at the present time, due to different basic priorities. If the drive of the four broad sections of the recycling industry, as described, could be drawn together, a comprehensive range of products could be specified with a price structure related to the degree of refinement rather than the method of production.

Local authorities should be allowed and encouraged to plan how to accommodate the following:

(a) The phased demolition of suitable buildings.
(b) Utilisation of the recycled demolition waste.
(c) The designation of areas suitable for recycling operations.

The implementation of more rigorous control and organisation would giving savings in terms of both time and costs and the UK's resources used to their full potential.

Plates

Plate 1 Type of precast concrete structures being demolished by the ball and chain method in Nottingham UK.

Plate 2 Common type of mobile crusher run equipment used in UK.

10. Table

Table shows sieve analysis required for a recycled crushed concrete aggregate.

Table 1.

DOT Granular sub base material type 1 (5)

BS sieve size	% by mass passing
75	100
37.5	85-100
10	40- 70
5	25- 45
600 um	8- 22
75 um	0- 10

determined by the washing and sieving method of BS812 Part 103.

Reference

1. Environmental Resources Limited (1980), Demolition Waste.
2. Trevorrow, A (1984) From Demolition to aggregate - Is going Dutch the answer? Building Technology and Management, Chartered Institute of Building.
3. Gallifords & Son, Junction 24 M1, 1987.
4. Sheriff Plant Hire, 1987, Colwick Industrial Estate, Nottingham.
5. Department of Transport. Specification for Highways Works, Part 3, Series 800 Road pavements. Unbound material. August 1986 H.M.S. Office.
6. Brown Crusher Hire Ltd., Leicester, January 1988.
7. Kue-Ken-Brown Lenox, Pontypridd, Glamorgan, Wales, November 1987.
8. Institution of Demolition Engineers.
9. London Waste Regulation Authority, County Hall, London 1986.

RECYCLED CEMENT AND RECYCLED CONCRETE IN JAPAN

K.YODA Kajima Corporation, Japan
T.YOSHIKANE and Y.NAKASHIMA Taiyu Kensetsu Co., Ltd, Japan
T.SOSHIRODA Shibaura Institute of Technology, Japan

Abstract
There are good ways to use waste-water sludge from ready-mixed concrete plants or demolished concrete with high cement content for "recycled cement". This report describes the results of experiments on the applications of two types of recycled cement, Type I and Type II, to concrete materials.

The experimental studies were carried out at two laboratories, Shibaura Institute of Technology (SIT) and the Central Laboratory of Taiyu Kensetsu (CLTK). The main experiments at SIT concerned the strength of concrete using recycled cement, or "recycled concrete." Different combinations were tested using concrete with Type I cement or ordinary portland cement and different methods of curing. The main experiments at CLTK concerned the effects of Type I and Type II cement on heat of hydration of concrete and the relationship between compressive strength of concrete and cement-water ratio.

As a result, data concerning the properties of recycled concrete were obtained. It was found that practical applications of recycled concrete could be to light structures, underground structures, and so on. In fact, ready-mixed recycled concrete using both recycled cement and recycled aggregate is already on the market in Japan.
Key words: Recycled cement, Recycled concrete, Sludge, Demolished concrete, Slag, Using recycled cement with ordinary portland cement, Curing method, Heat of hydration

1. Introduction

There has been a large amout of construction work in recent years; many urban areas, including the urban infrastructures, areas are being redeveloped.

Up to about a dosen years ago, sludge and demolished concrete could be disposed of easily for land reclaimation. However, in recent years there has been little land reclaimation, and the disposal of sludge and demolished concrete has become a big problem from the point of view of environmental conservation.

In this way, the reuse of waste materials has become a pressing necessity. Although standards have been established for water containing sludge to be used at plants for ready-mixed concrete, the quantity that can be disposed of in this manner is very small.

Part of demolished concrete can be broken up and reused as concrete aggregate or for the broken stone foundation of roads.

This report desclibes the main results obtained from various experiments on recycled cement.

Type I recycled cement largely consists of sludge, and Type II cement is made from demolished concrete with a high cement content. Both types of recycled cement are manufactured by Taiyu Kensetsu.

The experiments were carried out at two laboratories. One set of experiments, Series 1, was performed at Shibaura Institute of Technology, and the other, Series 2, was performed at the Central Laboratory of Taiyu Kensetsu. Series 1 consisted of experiments to determine the basic properties of concrete utilizing Type I cement, and the Series 2 experiments mainly concerned the application of both types of cement.

Recycled concrete is already used in the form of ready-mixed concrete for non-stress-bearing structures or low breast wall, and so on. The quantity shipped is 1,500-2,000m^3 per month. The price of recycled concrete is about 4% cheaper than normal concrete.

2. Out-lines of Recycled cements

2.1 Type I recycled cement

1)Main components: Sludge, blast-furnace slag, gypsum (2-3%), inorganic hydration accelerator.

2)Characteristics: The sludge needs to be dried at a high temperature (about 200 ^{o}C). This is because ettringite is a crystalline product in cement sludge and it changes into an amorphous substance during high-temperature dehydration.

2.2 Type II recycled cement

1)Main components: Crushed concrete with high cement content, blast-furnace slag, gypsum (2-3%), inorganic hydration accelerator

2)Characteristics: Crushed-concrete sand is made from demolished concrete with high cement content. Initially, concrete is crushed to particles of diameter 30mm, and then it is screened with a 5mm mesh. Compared with Type I, the quality is more stable, and less energy is required for drying.

3. Series 1 Experiments (at SIT)

3.1 Materials

1)Cement: Table 1 shows the physical properties of the different cements used, and Table 2 shows their chemical composition.

2)Aggregate: Table 3 shows the physical properties of the different aggregate used.

3)Water: Ordinary city tap water

3.2 Mix Proportion

Table 4 shows the mix proportions of the concrete.

Table 1. Physical properties of cement

Series	Type of cement	Spec. grav.	Spec. surf. (cm^2/g)	Setting time initial set. (hr-min)	final set. (hr-min)	Flow (mm)	Bend. strength 3d (kgf/cm^2)	7d	28d	Comp. strength 3d (kgf/cm^2)	7d	28d	91d
1	Ordinary Portland	3.14	3340	2-34	4-41	261	37	53	70	165	265	417	-
	Recycled Type I	2.79	4560	4-25	6-45	230	15	31	55	71	138	240	-
	Type C Portland blast-furnace	2.99	3970	3-24	4-48	250	18	34	69	69	141	365	-
2	Recycled Type I	2.77	6550	1-02	8-54	205	-	-	-	59	138	265	326
	Recycled Type II	2.83	5370	3-48	7-08	220	-	-	-	68	133	252	316

Table 2. Chemical composition (%)

Series	Type of cement	ig.loss[1)]	insol[2)]	SiO_2	Al_2O_3	CaO	Fe_2O_3	MgO	SO_3	Na_2O	K_2O	Total
1	Ordinary Portland	0.6	0.1	21.8	5.1	64.3	2.8	1.8	2.0	-	-	98.5
	Recycled Type I	1.6	16.1	23.0	8.5	40.6	2.0	3.8	2.1	0.3	0.6	98.6
	Type C Portland blast-furnace	0.8	0.5	28.4	11.2	49.8	1.5	4.0	2.2	-	-	98.4
2	Recycled Type I	1.9	-	29.2	10.6	49.4	1.6	4.2	2.4	-	-	99.3
	Recycled Type II	2.5	-	32.4	13.0	40.3	2.9	5.0	2.2	-	-	98.3

*1) ig.loss : ignition loss, 2) insol : insoluble ratio

Table 3. Physical properties of aggregate

Series	Aggregate	Type	Maximum size (mm)	Specific gravity when dry	Absorption (%)	Unit weight (kg/ℓ)	Percentage of absolute volume (%)	Fineness modulus
1	Fine	River sand	5	2.28	6.10	1.50	65.6	2.84
	Coarse	River gravel	20	2.49	2.37	1.59	65.4	6.90
		River gravel	20	2.41	3.89	1.44	62.1	6.80
2	Fine	River sand	5	2.53	1.6	1.61	-	2.72
	Coarse	River gravel	25	2.59	0.8	1.64	63.3	6.83

Table 4. Mix data of concrete (S.I.T.)

No.	Mix	Cement	Coarse aggregate	W/C (%wt)	W/C (%vol)	s/a (%vol)	Water content (kg/m³)	Combined proportion of cement (%) Recycled	Combined proportion of cement (%) Ordinary Portland
1-1	CN-N	CN	N	70	221	49.3	208	0	100
	CN-R	CN	R					0	100
	C100R-R	CR	R	77				100	0
1-2	CN	CN	N	60	188	34.6	196	0	100
	C25R	Mix	N		183	34.3		25	75
	C50R	Mix	N		180	34.2		50	50
	C75R	Mix	N		175	33.9		75	25
	C100R	CR	N		172	33.6		100	0

Cement CN: Ordinary Portland Cement   Coarse aggregate N: Normal

CR: Recycled Cement Type I   R: Recycled

3.3 Results and Discussion

1)Fresh concrete: Recycled concrete does not segregate easily because of its high specific surface. (The specific surface of Type I cement is 4,560cm^2/g, while that of Type C portland blast-furnace slag cement is 3,970cm^2/g).

2)Strength

i) Variation of cement and aggregates (fixed wet curing)

As can be seen from Fig. 1, concrete with recycled cement and recycled aggregate (C100R-R) has about half the strength of normal concrete with ordinary portland cement (CN-R).

The decrease in strength is caused by the different cement rather than the different aggregate.

If the cement is same and the aggregate is varied, Fig. 1 shows that concrete with recycled aggregate (CN-R) has an 8.5% higher strength than normal concrete (CN-N).

This is caused by the fact that normal aggregate has poor qualities such as low oven-dry specific gravity (2.49) and high percentage of water absorption (2.37%). Moreover, recycled aggregate has few impurities.

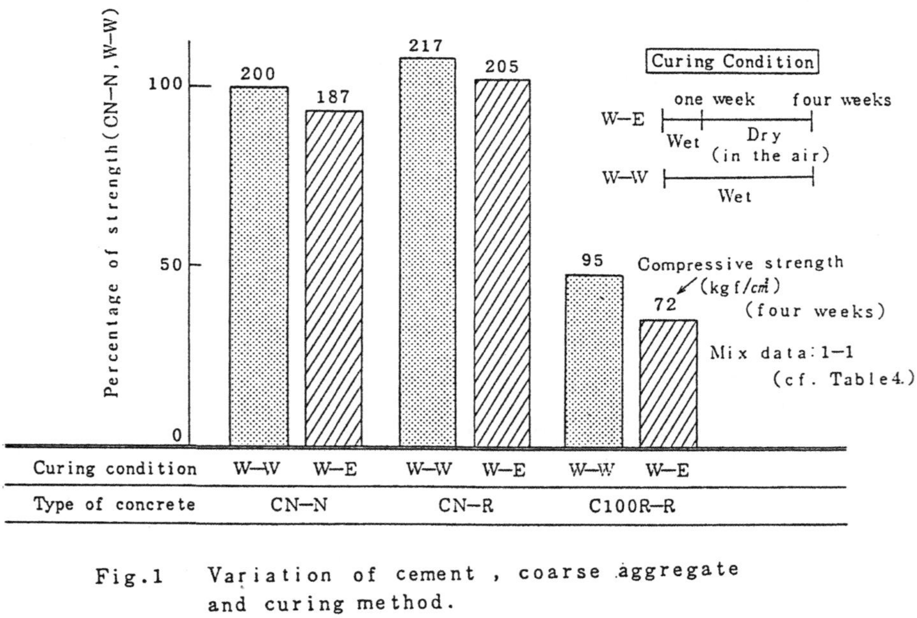

Fig.1 Variation of cement, coarse aggregate and curing method.

ii) Variation of ratio of Type I cement to ordinary portland cement. As can been seen from Fig. 2, with 4 weeks as the line of demarcation, the curve for 1 week is convex and for 13 weeks and over, the curves are concave. Type I recycled cement can replace ordinary portland cement in concrete with no decrease in strength if the ratio of Type I cement is 25% to 75% ordinary portland cement. (provided the concrete is wet cured for a long time). Recycled cement contains blast-furnace slag which is affected by the ordinary portland cement as an accelerator, so that its latent hydraulic property gives it a higher strength.

iii) Variation of method of curing

As can be seen in Fig. 1, For all types of concrete, W-W curing (remaining in water) resulted in a strength of 6-32% higher than W-E curing (first 1 week wet curing, then dry curing). Therefore, the best method of curing is wet curing. The difference is particularly large for concrete using recycled cement because of the latent hydraulicity of the slag it contains.

Compressive strength of concrete with ordinary portland cement.

Age (weeks)	Strength (kgf/cm²)
1	212
4	303
13	354
26	366
52	364

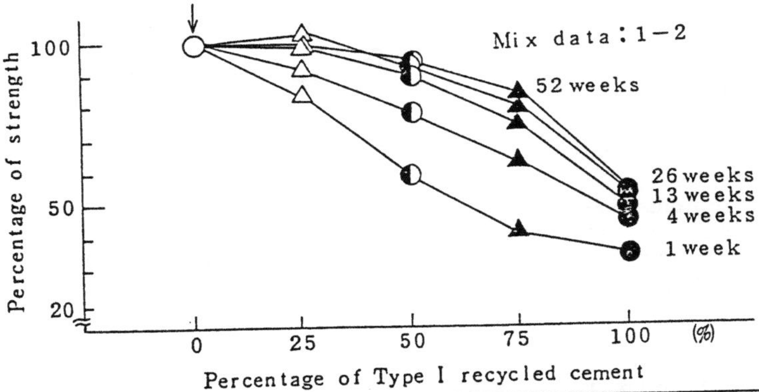

Mark :	CN	C25R	C50R	C75R	C100R

Fig. 2 Variation of ratio of Type I cement to ordinary portland cement

3)Drying shrinkage (of mortar)
As can be seen in Fig. 3, drying shrinkage of mortar does not vary much except for C100R recycled cement can replace ordinary portland cement in mortar without dramatically increasing the drying shrinkage provided that the ratio of recycled cement is low (25%).This is especially true if the mortar is wet cured for a long period, which is the same tendency as for strength.
4)Dynamic modulas of elasticity
The tendency for dynamic modulus of elasticity of the concrete is almost the same as for strength. Recycled cement can replace ordinary portland cement in concrete up to the ratio 50% without significantly changing dynamic modulus of elasticity, and this is particularly true if the concrete is cured for a period.
5)Carbonation
For different types of concrete, carbonation happens fastest when the cement is 100% recycled, followed by Type C portland blast-furnace slag cement, 50% recycled cement to 50% ordinary portland cement, and finally ordinary portland cement.

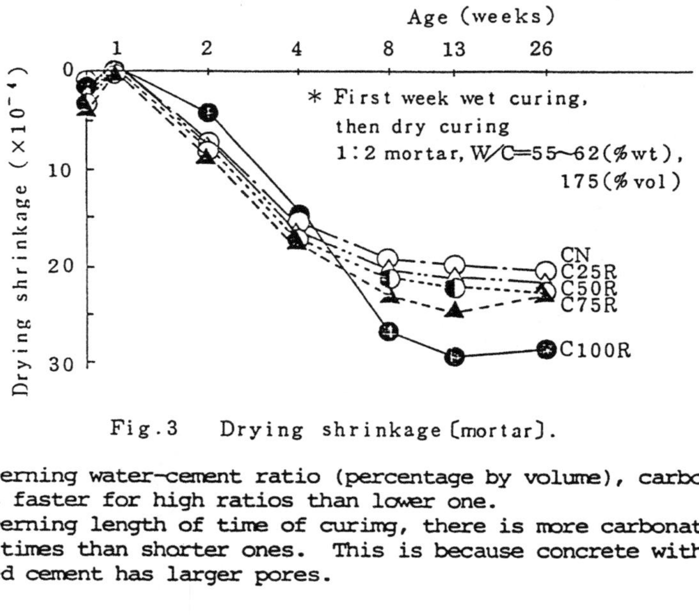

Fig.3 Drying shrinkage [mortar].

Concerning water-cement ratio (percentage by volume), carbonation happens faster for high ratios than lower one.

Concerning length of time of curimg, there is more carbonation for longer times than shorter ones. This is because concrete with recycled cement has larger pores.

4. Series 2 Experiments (at CLTK)

4.1 Materials
1)Cement: Table 1 shows the physical properties of the different cements used, and Table 2 shows their chemical composition.
2)Aggregate:
Table 3 shows physical properties of the different aggregates used.
3)Water: Ordinary city tap water

4.2 Mix Proportion
Table 5 shows the mix proportions of the concrete.

4.3 Results and Discussion
1)The relationships between cement type and sand-to-aggregate ratio (s/a), and water-to-cement ratio (W/C).
To maintain same workability and compressive strength, the ratio s/a for recycled concrete with Type I or Type II recycled cement should be 5-6% smaller than for concrete with Type B portland blast-furnace slag cement; the ratio W/C should be respectively 1-2% larger. (slump: 15cm, compressive strengh: 180kgf/cm^2-28days)
2)The relationships between water-to-cement ratio and compressive strength.
As can be seen in Fig. 4, after 4 weeks of wet curing, the compressive strengths of concrete with Type I and Type II recycled cement are almost the same, but they are both lower than for Type B portland blast-furnace slag cement. However, after wet curing for 13 weeks, the compressive of the three different type of concrete almost the same.
3)Drying Shrinkage
The drying shrinkages for all three types of concrete and for any method of curing do not vary significantly.
4)Freezing / thawing durability
The freezing / thawing durabilities of all three types of concrete do not vary significantly.
5)Heat of Hydration
As can be seen in Fig. 5, the heat of hydration of the both types of concrete with recycled cement is lower than that of Moderate-heat portland cement.
6)Carbonation
Carbonation of both recycled concrete is faster than concrete with of Type B portland blast-furnace slag cement.

Table 5. Mix data of concrete (CLTK)

No.	Cement	W/C (%wt)	s/a (%vol)	Mix proportions of concrete (kg/m^3) C	W	S	G	Additive AE & WR
2-1	Recycled Type I	61	40	279	170	707	1078	0.558
	Recycled Type II	60	41	280	168	727	1067	0.560
	Type B Portland blast-furnace slag	68	46	246	167	840	1002	0.492

* W/C : water-to-cement ratio
s/a : sand-to-aggregate ratio
C : cement, W : water
S : fine aggregate, G : coarse aggregate

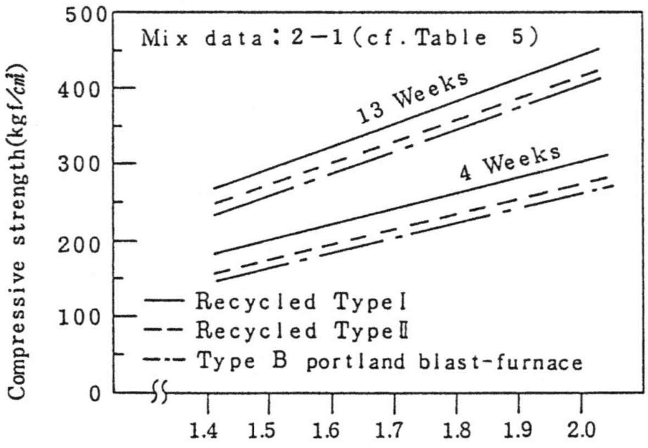

Fig.4 Relationship between cement to water ratio and compressive strength

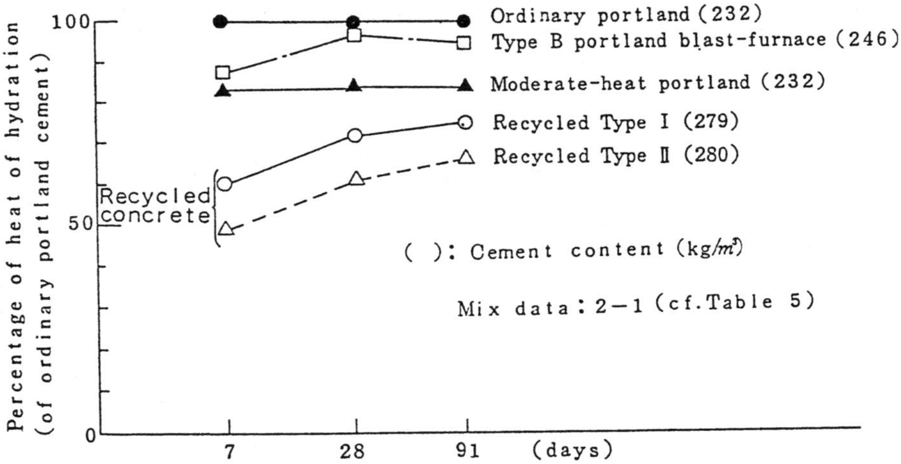

Fig.5 Variation of heat of hydration with type of cement

5. Conclusions

The main results obtained were as follows:

1)Characteristics of cement

Comparing Type I and Type II recycled cement to Type C portland blast-furnace slag cement, the former have higher specific surface, more insoluble residue, lower specific gravity, less CaO content.

2)Properties of fresh concrete

The slump and flow of concrete using recycled cement are lower than those of concrete with ordinary portland cement, provided the mix proportions by volume are the same, allowing slump and strength to vary. This is because the differences in specific surface and specific gravity affect the capacity to hold water and viscocity.

3)Characteristics of concrete with recycled cement

After wet curing for 4 weeks, the compressive strength of concrete with recycled cement is half that of concrete with ordinary portland cement. The trends in the other properties of concrete with recycled cement are almost the same as for compressive strength.

To improve the quality of concrete with recycled cement, the following are necessary: (1)Thorough wet curing, (2)Low water-to-cement ratio, (3)To use of ordinary portland cement as an accelerator.

Comparing concrete using recycled cement with Type B portland blast-furnace slag cement, if compressive strength and slump are fixed, development of strength, drying shrinkage and freezing/thawing durability are almost same, while heat of hydration is lower, and carbonation is faster.

These characteristics of recycled concrete with recycled cement are suitable for "light grade" concrete, foundations, piles, and mass concrete. As for using it for pavements, curing and finishing of the surface may be required more.

Regarding future directions of research, new manufacturing techniques for lower cost cement will be investigated, utilizing the characteristics of low heat of hydration and resistance to segregation. The development of higher quality recycled cement will be attempted.

Acknowledgements

The authors are very grateful to those concerned with the experimental works in both laboratories.

References

Yoda,K. and Soshiroda,T. (1986) Reuse of sludge for concrete. Trans. Japan Concrete Institute, No.8, 865-868.

Yoshikane,T., Takeshima,H. and Nakashima,Y. (1986) Recycled cement made from waste concrete and its application to concrete. Trans. Japan Concrete Institute, No.8, 861-864.

Yoda,K. and Soshiroda,T. (1987) Recycled concrete with recycled cement (Part III.) Summaries of technical papers of annual Meeting, Architectual Institute of Japan, 27-28.

ON-SITE USE OF REGENERATED DEMOLITION DEBRIS

JENS BJØRN JAKOBSEN
MORTEN ELLE
Department of Environment
COWIconsult, Consulting Engineers and Planners AS
ERIK K. LAURITZEN
Demex, Consulting Engineers AS

Abstract

The generation of solid waste from construction and demolition sites in Denmark is approximately 1.25 million tons/year. Recovery of materials from this type of waste can reduce landfill areas and consumption of gravel resources. Obstacles for an extended recovery exist. A demonstration project is, therefore, conducted to prove the feasibility of on site use of regenerated demolition debris.

An old factory is demolished. The total waste quantity is 8,600 tons. Approximately 3,400 tons of tile and concrete is crushed by a mobile crushing plant. The crushed materials are used for constructing a parking area next to the demolished factory. The crushed materials are analyzed (sieve analysis, density, etc.) at a laboratory. Five different types of pavements are established. The pavements are tested during construction (plate loading, standard proctor, etc.). Laboratory tests show acceptable quality of the crushed materials. The possibilities for on site utilization is verified by the site tests.

Key words: Demolition debris, Crushed materials, Reuse, Regeneration.

1. Introduction

In 1984 the generation of solid waste from construction and demolition sites in Denmark was estimated to 1.25 million tons/year. This corresponds to 250 kg/inh/year. In addition 0.75 million tons was generated at civil works corresponding to 150 kg/inh/year.

The highest portion of waste was generated at demolition sites, rising to 0.6 million tons/year (120 kg/inh/year). Highest quantity of demolition waste was originated from demolished factories.

Surveys have been made by Cowiconsult and Demex on the quantity and compostion of demolition waste. The specific generation of waste from demolition sites arises to 0.9 - 1.1 tons per square meter building. The composition of the waste depends very much on the type and age of the building. In Denmark multistorey dwellings, from the turn of the century, consist of 80% bricks of tile/concrete, 15% wood/timber and 5% miscellaneous.

There are a number of advantages in recovering materials from demolition, as well as, construction works. Among others, in evidence, is the reduction of landfill area and consumption of gravel resources. However, in spite of these advantages there are some obstacles in the extended recovery of materials from demolition and construction waste. These obstacles can change from region to region and from country to country.

Typical problems are:

- Virgin resources of gravel are inexpensive and easily accessible.
- Disposal of waste on landfill is inexpensive and without any limitations.
- The demolition works are not designed for materials recovery, giving mixed waste materials not suitable for any utilization.
- The need for technical guidelines or standards for utilization of recovered demolition waste.

Recovery of materials will not take place without overcoming one or more of the above mentioned obstacles.

2. Advantages in the Recovery of Demolition and Construction Waste.

A number of measures can be taken for increasing the materials recovery from demolition and construction works. These measures will depend on the final utilization of the materials.

Materials from demolition and construction works can be recovered within a geographical region (country, etc.) or within a very short distance from the waste production site (local recovery).

The overall advantages are mentioned above. Local recovery gives, in addition, the following advantages:

- Reduced external transport of waste to landfill sites (i.e. reduction of traffic).
- Lower risk of mixed and non-controlled waste materials, producing greater probability of high quality, upgraded materials for later utilization.
- Reduced transport of gravel to sites for new construction.

There will, however, be a possibility of negative impact to the surroundings from noise, dust and vibrations from the crushing and sorting of the waste.

Owing to the interesting perspectives within local recovery of demolition waste, it was decided to carry out a demonstration on local recovery.

3. Objectives of Demonstration Project

The overall objectives of the demonstration project were defined as:

- To design and survey a selected demolition project from bricks of tile and concrete, which were to be upgraded to gravel.
- To upgrade (crush and sort) the sorted bricks from the demolition.
- To test the produced gravel.
- To utilize the gravel for constructing a parking area.
- To test the quality of the upgraded materials within the mentioned utilization.

The demonstration project was designed and supervised by COWIconsult and Demex, both consulting engineering companies. Financing was made by the Danish Agency of Environmental Protection and by the Danish Council of Technology.

4. Demonstration Project - Demolition

An old factory building was selected as the object for the demonstration demolition. The factory represents a typical old industrial building comprised of both tile and concrete materials, as well as reasonable quantities of timber. Further, the factory was located adjacent to a theater with parking space deficiencies.

Specifications for the factory building are as follows:

- Building area	approx. 3,400 m^2
- Storey area	approx. 7,800 m^2
- Waste quantity	approx. 8,600 tons
. Tile	approx. 3,500 tons
. Mixed demolition waste	approx. 5,000 tons
. Wood	approx. 70 tons
. Steel	approx. 470 tons

The layout plan for the factory and surrounding areas is shown below.

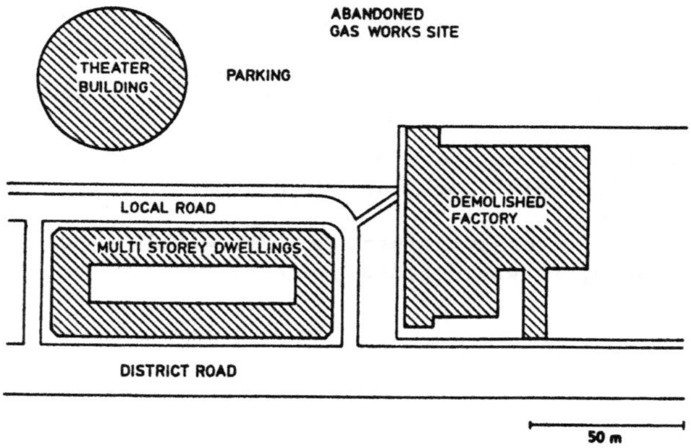

The time period for the demolition work was 6 weeks. Steel as well as good quality timber was sorted out separately. Tile walls, concrete storey floors and roof constructions were demolished as individually as possible.

During the demolition process an attempt was made to segregate bricks of tile and concrete from wood, steel and other materials. Owing to the tight time limit, for the contractor, a 100% efficient segregation was not possible.

Approximately 3,400 tons of tile and concrete waste was sorted out for later crushing. The residuals (5,000 tons) was disposed of on a landfill.

5. Demonstration Project - Crushing of Demolition Waste

5.1 Crushing Capacity

The crushing of the sorted tile and concrete was carried out by a mobile crushing plant, "Combi Screen" produced by "Vedbysønder Maskinfabrik". The crushing plant was erected on the adjacent theater ground. The erection time was 3 days.

Specifications for the crushing plant are as follows:

- Feeder:	volume	10 m^3
	capacity	180 m^3/h
- Scalper:	two storeys	50 mm
- Crusher:	type Hazemag AP-S1313. Impact	
	. capacity 120 - 180 tons/h	
	. max. size 0.5 x 1.0 m	
	. power 150 hp	

Dust from the crushing operation is cleaned through a cyclon precipitator.

The crushing plant had a two man crew. The waste was fed to the plant by means of a hydraulic shovel periodically assisted by a dozer.

During the crushing a number of stops and breakdowns occured. The reason for the stops varied from problems with the belt conveyors to problems with screens, crusher, etc.

The total time of operation for the crushing plant was 152 hours. Running time for the crusher was 70 hours giving an output of 49 tons of demolition waste per hour.

A total of 300 tons of demolition waste was crushed, comprising:

- . 250 tons of concrete (non reinforced),
- . 300 tons of mixed concrete and tile from foundation of the buildings,
- . 3,350 tons of tile.

5.2 Measurements of Noise

Noise level, in the surrounding area, was measured during the crushing. Valid information was registered about noise levels from the crushing plant for use in other similar projects.

The primary source of noise was the diesel engine (power generator) crusher, vibrating screen and pay loader. The noise level was measured twice at five different positions as shown on the sketch below.

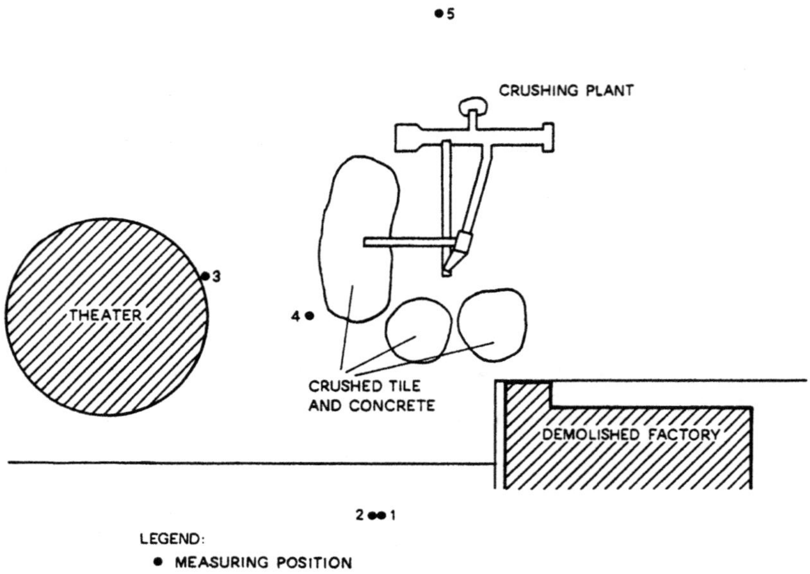

All measurements took place at full scale operation of the plant. The second measurement at positions 3 and 4 were blocked by heaps of crushed materials. Position 1 was partially blocked.

The results from the measurements are shown in the table below.

Noise Level L_{Aeq} Position	1	2	3	4	5
Unblocked	71	74	73	74	76
Blocked	69	-	64	63	-

Based on the measurements above the mean A-weighted noise level is calculated to $L_{WA} = 117$ dB. The mean A-weighted noise spectrum per 1/1 octave is shown in the table below

Hz	63	125	250	500	1k	2k	4k	8k
dB(A)	102	104	107	109	112	111	106	97

5.3 Quality of the Crushed Materials

The quality of the crushed materials was examined by the Danish National Road Laboratory. Samples were taken from three types of crushed materials: tile, mixed tile and concrete as well as concrete.

The results from the laboratory analysis are shown in the table below.

Analysis no.		Tile		Mixed tile/concrete	Concrete
Parameter		1	2	3	4
Uniformity		52	47	42	7.5
Screened >16 mm(s)	(%)	20	18	21	34
Sand equivalent (SE)	(%)	69	49	67	70
Capillarity (h_c)	(cm)	35	50	30	20
ASTM d,max	(t/m^3)	1.68	1.75	1.79	1.79
Water content (W_{opt})	(%)	19.0	17.0	15.7	11.2
Modified proctor					
· d,max	(t/m^3)	1.68	1.69	1.74	1.79
· W_{opt}	(%)	17	13	14.5	13.5
Specific gravity (0,075-4mm)	(t/m^3)	2.57	2.59	2.60	2.57
Absorption (0,075-4mm) W_a	(%)	8.3	8.3	9.5	6.8
Specific Gravity (4-32mm)	(t/m^3)	2.45	2.59	2.58	2.57
Absorption (4-32mm)W_a	(%)	13.5	13.0	8.9	4.4
Los Angeles test	(%)	40	41	37	29
Light particles [1] (4-32mm)<2,2t/m^3)	(%)	-	27.1	19.2	1.5
Light particles [2] <1,0 t/m^3	(%)	-	0.3	0.0	0.0

[1] Measured in a ZuBr solution.

[2] Measured in water.

Below is shown the sieve analysis for tile (1) and concrete (4).

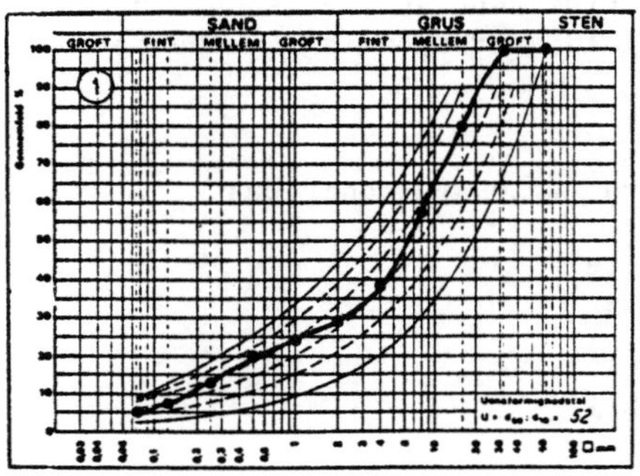

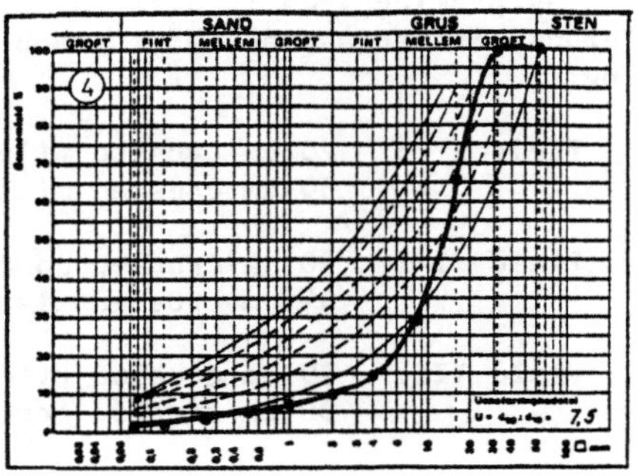

The curve for the crushed tile is comparable to that of high quality gravel. The curve for crushed concrete shows a quality having a too high content of medium and coarse gravel.

Since all crushing was done with the same settings on the crushing plant this difference must originate from the components of the materials. The high content of medium and coarse particles in the crushed concrete probably reflects the original aggregate gradation.

There is a risk that material with these characteristics may prove unstable under loading, whereas the crushed tile and maxed products should present no such problem.

6. Construction of a Parking Area from Crushed Tile and Concrete

As mentioned earlier, the crushed materials from the demolished factory form the basis for the construction of a parking area on space next to the factory site. The parking area would be used for spectators visiting the theater neighbouring the factory.

The purpose of this part of the project is to assess the crushed materials as sub-bases and unbound layers within low priority utilization. The parking area will not be paved with asphalt.

A total of approximately 1,000 m^2, corresponding to 42 parking slots, are established as shown on the sketch below:

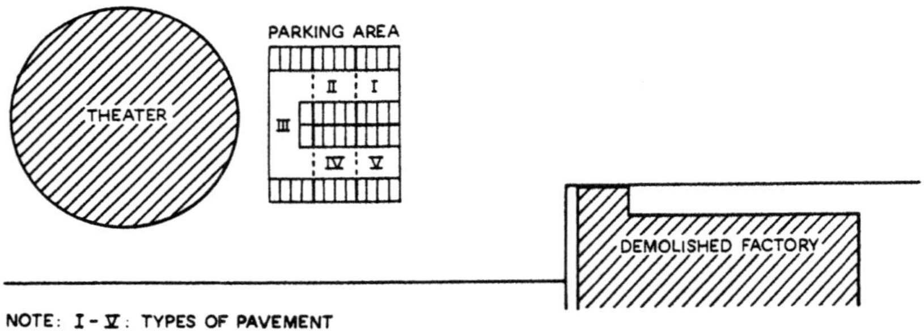

The parking area is divided into five separate sections. Different pavements are used in each field corresponding to the following:

Field I :	20 cm crushed tile 30 cm crushed mixture (tile and concrete)
Field II :	20 cm crushed concrete 30 cm crushed mixture
Field III:	20 cm crushed tile 30 cm crushed concrete

Field IV : 20 cm crushed tile
30 cm crushed concrete

Field V : 20 cm crushed tile
30 cm crushed tile

During the execution of the parking area, selected parameters are measured. The intention of these measurements are to establish for each layer the:

- deformation characteristics,
- obtainable density,
- compaction requirements for standardized density.

The programs for the measurements, etc. comprise the following:

Sub-base
a Plate loading test on foundation level,
b Compaction by a vibratory roller (8 passes) to 95 % standard proctor,
c Nuclear densimeter respective sandfilling tests in situ and laboratory determination of Standard Proctor density.

Base Layer
d Compaction by a vibratory roller (12 passes) to 100% standard proctor,
e Measurement of density as indicated for sub-base layer,
f Falling weight deflectometer (FWD) test to determine E-moduli of sub-base and base layers.

The results from the measurement c and e are given in the table below.

		Density (Standard Proctor)	
		Nuclear Densimeter %	Sandfilling %
Sub-base	Concrete	96.7/98.7	99.0/103.4
	Mixture	93.2/	95.0/-
Base Layer	Tile	94.2/92.5	94.3/91.8

The density of the tile base was found to be far below the 100% Standard Proctor requirement, which was the ultimate goal. The tile base was therefore compacted with 12 more passes. The density thereafter reached an average of 95.6% (nuclear densimeter) corresponding to 98.3% standard proctor.

As seen in the table, the compaction effort is higher for tile than for concrete. The tile, however, has properties which makes it as suitable as concrete for the actual project.

At the time of the finalization of the written paper, the E-moduli had not been established by the mentioned FWD-tests. The results will be given in the oral presentation of the paper.

The measurements mentioned will be followed up by long term performance measurements (three years) to invesitgate the possible stability problems with the crushed concrete and to monitor any changes in bearing capacity. It is also planned to check the extent of degradation that will be caused by the effects of traffic and weather.

7. Conclusion and Long Term Visions

During the demolition works, approximately 3400 tons of demolition debris was sorted out for local utilization. A total number of 170 heavily loaded trucks thereby avoid the local and regional road system. This corresponds to a reduction by one third of the traffic from the demolition site.

Further, the utilization of crushed materials for the parking area instead of gravel producrs a savings of 170 or more trucks fully loaded with gravel.

Finally, the reuse of the regenerated demolition debris leads to savings of approximately 3,500 m^3 landfill volume and approximately 2,100 m^3 in the use of gravel.

The crushing operations, however, caused a noise level in the surrounding area, which was noticeably higher than the normal daily noise level. No complaints were, however, received.

Furthermore, no problems occured from dust and vibrations. Dust precautions were taken both by means of a dust precipitator connected to the crushing unit and by means of sprinkling the crushed materials during heaping.

The capacity of the crushing plan was only 30-40% of the nominal capacity specified by the supplier. The reduced capacity caused very high crushing cost. The cost per ton of crushed materials is calculated to approximately 90 kr./ton (14 US$ per ton) which is approximately twice the price of the gravel delivered to the demolition site.

The crushed concrete and tile were of comparatively high quality. Even though segregation at the demolition site was limited, the content of wood in the tile was only 0.3%. The analyzed charateristical parameters shows that the crushed materials - with few exceptions - are comparable to that of virgin gravel.

The concrete shows, as expected, the best parameters (except for uniformity), but the crushed tile is better than anticipated.

Measurements during the construction of the parking area shows that both crushed concrete and tile can fulfill traditional requirements for the specified purpose. Long term performance measurements will show if the crushed concrete/tile can fully substitute gravel in this context.

Local reuse of regenerated demolition debris will, in the future, depend very much on the local conditions. These are among others:

- Size of demolition work,
- Demand for and value of gravel substitution in the neighbourhood,
- Distance to nearest regional crushing plant/regional landfill,
- Price relations between gravel, crushing costs and landfill, taxes.

It is estimated, that only demolition projects producing more than 20,000 tons are suitable for local crushing. Local utilization of the crushed materials will only take place if new construction, at the same place (or next to), demands approximately the same quantity of gravel substitution. Local utilization further requires time adaption for the new construction as well as storage facilities for the crushed materials.

In long term visions only few demolition works are available for genuine local reuse.

PROPERTIES AND USAGE OF RECYCLED AGGREGATE CONCRETE

SHIGETOSHI KOBAYASHI and HIROTAKA KAWANO
Public Works Research Institute, Ministry of Construction
Japanese Government

Abstract
The Ministry of Construction of the Japanese Government carried out research into the utilization of wastes from construction works, and the development of reuse techniques for demolished concrete was one of the main subjects. After some feasibility studies, recycled aggregate was taken up, as a coarse and fine aggregate extracted mechanically from demolished concrete. The results of laboratory-experiments for the properties of recycled aggregate and concrete with recycled aggregate, showed that the specific gravity and absorption were affected by the amount of cement paste which was sticking on the aggregate and as the soundness of aggregate was related to absorption, the properties of concrete with recycled aggregate with high absorption tended to be worse in strength and in resistance to freezing and thawing than those with ordinary aggregate. The field tests showed no problems in the mixing, placing, and compaction procedures. Some recommendable usages for concrete with recycled aggregate were also proposed by considering the experimental results and expenses for refinning the aggregate.
Key Words: Recycled aggregate, Recycled aggregate concrete, Property, Durability, Recycle, Aggregate, Concrete, Strength

1. Introduction

The Annual production of ready-mixed concrete in Japan is 160 million m^3. Even if we are conservative by assuming that 10% of the annually produced concrete is subjected to demolition, a huge amount of waste concrete is generated, resulting in a serious problem in acquiring disposal sites. On the other hand, the waste concrete is hardly utilized as recycled aggregate for concrete. The major reasons for this seems to be that the aggregate taken from crushed concrete cannot meet the quality standards for general aggregate for concrete because this aggregate contains a lot of cement paste with a resultant high water absorption rate, and no adequate utilization guideline for the recycled aggregate are available, even if one wants to treat it separately. Recently, however, a machine which removes the paste sticking to the aggregate has been developed, and may allow the manufacturing of good quality concrete from recycled aggregate. In response to this trend, the Ministry of Construction has established a research project to investigate the performance of recycled aggregate and the quality of concrete utilizing it. The results from this project are summarized below.

2. The quality of recycled aggregate

The removal of cement paste adhering recycled aggregate by mechanical crushing and scrubbing has been put to practical use, and these techniques have been further refined. The recycled aggregate used for this study was made by a recycled aggregate manufacturing plant using a refining process developed by KEIHAN Concrete Company in Kyoto.

Concrete specimens of two different strengths, approximately 40N/mm^2 as the higher strength and approximately 20N/mm^2 as the lower strength, were made using same the original aggregate in order to determine the strength of the concrete from which the aggregate was taken (original concrete). These specimens were crushed to make the recycled aggregate for this study. The original aggregate was composed of crushed stone and pit sand. Table 1 summarizes the types of aggregates used for this study. The recycled aggregates carry the mark M, while numerical values following the M indicate the degree of refining (the removing of cement paste) treatment, the greater the number, the higher the degree of refinement. Table 2 and 3 show the test results for each aggregate.

Table 1 Types of aggregates

Types		Strength of original concrete		Notes
		Low strength	High strength	
Fine aggregate	Pit sand	—	—	Fine aggregate of the original concrete
	MS-1	○	○	Less than 10mm material crushed by an impact crusher and then treated once by a fine aggregate refining machine.
	MS-3	○	○	Less than 10mm material crushed by impact crusher and then treated three times by a fine aggregate refining machine.
Coarse aggregate	Crushed sand	—	—	Coarse aggregate of the original concrete.
	MG-0	○	○	Only crushed by impact crusher.
	MG-1	○	○	MG-0 refined only with an impact of a coarse aggregate refining machine.
	MG-5	○	○	MG-0 refined by an impact, destructive power and scrubbing exerted by a coarse aggregate refining machine.

Table 2 Physical properties of recycled coarse aggregates

Types	Strength of original concrete 4)	Maximum size (mm)	Specific gravity 1)	Absorption (%) 1)	Soundness, lost mass (%)	Abrasion loss (%)	Bulk density (kg/l)	Solid volume percentage (%)	Fineness modulus	Solid volume percentage for shape determination (%) 3)
MG-0	High	25	2.49 2) 2.48	3.98 4.02	22.7	26.6	1.42 1.41	59.2	7.06	-
	Low	25	2.34 2.38	4.78 5.18	31.5	28.6	1.33 1.34	59.3	7.12	57.6
MG-1	High	20	2.56 2.57	2.53 2.48	6.4	20.0	1.54 1.54	61.4	7.04	-
	Low	20	2.55 2.55	2.28 2.56	17.5	19.8	1.51 1.52	61.0	6.92	59.5
MG-5	High	20	2.60 2.62	1.59 1.55	9.6	11.9	1.60 1.62	62.8	6.88	-
	Low	20	2.65 2.66	1.09 1.06	5.8	14.6	1.62 1.64	62.4	6.82	60.4
Ibaraki crushed stone	-	25	2.69	0.70	6.2	13.7	1.56	58.4	7.07	58.3

(Notes) 1) Specific gravity and absorption are obtained after soaking 24 hours in water.
2) Figures written in two rows are values measured in two tests.
3) Obtained by the method of JIS A 5005 "Crushed Stones for Concrete".
4) High strength and low strength mean approximately $40N/mm^2$ and $20N/mm^2$.

Table 3 Physical properties of fine aggregates

Types	Strength of original concrete 3)	Specific gravity 1)	Absorption (%) 1)	Soundness, lost mass (%)	Bulk density (kg/l)	Solid volume percentage (%)	Fineness modulus
MS-1	High	2.36 2) 2.37	7.76 7.84	2.1	1.43	65.0	2.76
	Low	2.39 2.41	6.62 6.58	4.8	1.45 1.45	64.2	3.16
MS-3	High	2.45 2.45	5.77 6.58	2.0	1.45	62.8	2.50
	Low	2.51 2.52	2.96 3.76	3.4	1.54 1.55	63.8	2.53
Jyouyo pit sand	-	2.56	1.79	1.8	1.65	65.7	2.75

(Notes) 1) Specific gravity and absorption are obtained after soaking 24 hours in water.
2) Figures written in two rows are values measured in two tests.
3) High strength and low strength mean approximately $40N/mm^2$ and $20N/mm^2$.

Differeces in the property between the recycled aggregate and the original aggregate seem to be due to the adhesion of cement paste to the aggregate. Figure 1 gives the relation between the measured amount of paste adhering to the recycled aggregate and the water absorption rate of the recycled aggregate.

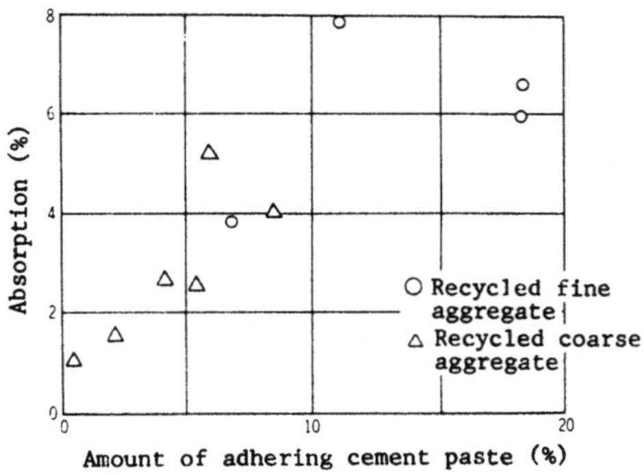

Figure 1 Amount of adhering cement paste and absorption of recycled aggregate

Since the water absorption rate of original aggregate is generally lower than that of the cement paste, a higher amount of adhering cement paste tends to result in a higher absorption for the recycled aggregate. In particular, in the case of recycled coarse aggregate, a linear relation is observed between the two.

The data show that the water absorption rate of MS-1 recycled from the high strength concrete is 7.84% or 4.4 times of the original fine aggregate, while the water absorption rate of MG-0 recycled from the low strength concrete is 7.1 times that of the original coarse aggregate. In addition, the water absorption rate of the recycled fine aggregate subjected to only primary crushing is 11.2% or approximately 9 times that of the original fine aggregate. However, a higher degree of refining can sharply reduce the water absorption rate of the recycled aggregate.

The recycled fine aggregate tends to have a higher absorption rate because the fine aggregate has a higher amount of adhering cement paste.

The specific gravities of the recycled fine aggregate and the recycled coarse aggregate are lower than those of the crushed stone and the mountain sand used for the original concrete (hereinafter refered to as "the original coarse agrregate" and the original fine aggregate, respectively), but the higher degree of refining leads to a specific gravity close to that of the original fine and coarse aggregates. In addition, the difference in properties of the recycled aggregates with different strengths for the original concrete are as follows. For MS-1 and MG-5, low strength concrete produces a higher specific gravity, while for MG-0, high strength concrete yields a lower specific gravity. This seems to be because the paste portion is easily removed by the recycling treatment, although a lower strength of the original concrete produces a lower specific gravity for the concrete.

Figure 2 shows the relationship between the specific gravity and the absorption. For both the fine and coarse aggregate, the absorption rate decrease linearly in proportion to the specific gravity.

The lost mass percentage of the recycled fine aggregate and that of the recycled coarse aggregate were 2.0 - 4.8%, and 11.9 - 28.6%, respectively, showing fairly different results between the fine aggregate and the coarse aggregate. The recycled coarse aggregate tends to have a lower lost mass percentage with a higher degree of refining. Both the recycled fine and coarse aggregates show larger lost mass with a strength of the original concretes is lower except MG5.

Figure 3 gives the relationship between the absorption and lost mass percentage. The relationship between these two parameters of the recycled fine aggregate is quite different from that of coarse aggregate.

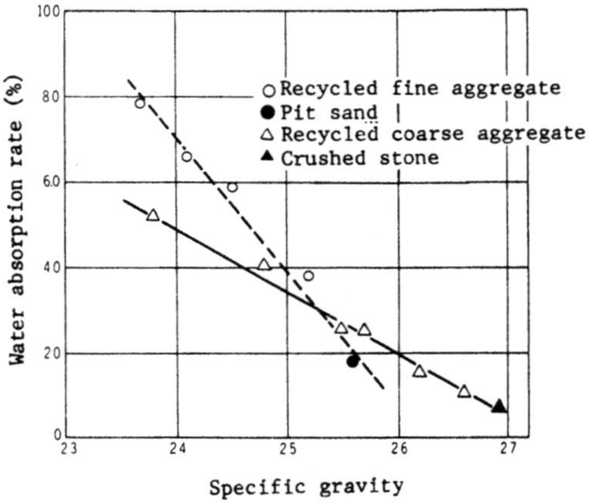

Figure 2 Relation between specific gravity and water absorption rate

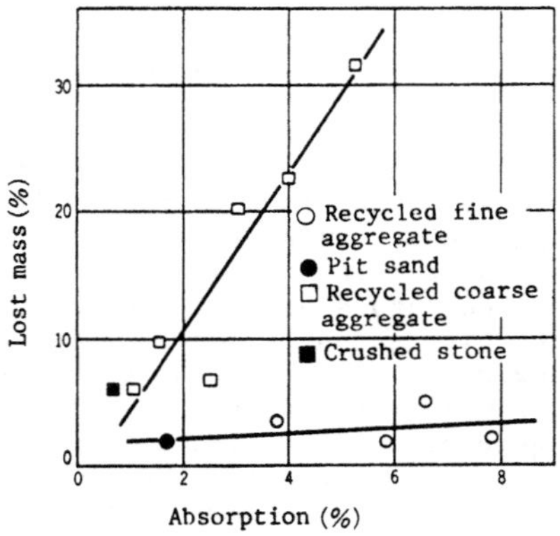

Figure 3 Water absorption rate and stability loss wight

The maximum abrasion loss of the recycled fine aggregate was 28.6%, but the loss was reduced to a value equivalent to or below that of the original coarse aggregate by adopting a higher degree of refining.

The bulk density and the solid volume percentage of the recycled fine aggregate were lower than those of the original fine aggregate. It is estimated that the initial recycling treatment of impact crushing produced angular recycled fine aggregate.

On the other hand, MG-5 of the recycled coarse aggregate had a greater bulk weight than the original coarse aggregate, with a higher solid volume percentage than the original coarse aggregate. As the solid volume percentage is affected by particle shape, it becomes higher with a higher degree of treatment. This is due to that a higher degree of the refining leads to stronger abrasion during the treatment process, resulting in a round shape for the aggregate.

Changes in the absorption rate of the recycled fine and coarse aggregates with time are shown in Figures 4 and 5. These show the prewetting time prior to the concrete mixing of the recycled aggregates should be taken at least 1 day.

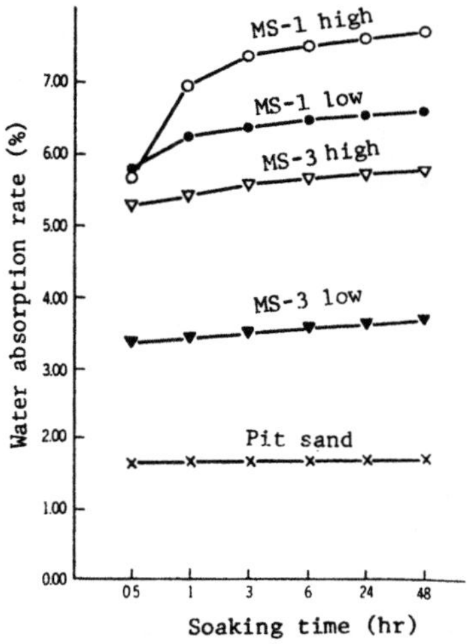

Figure 4 Soaking time and absorption for recycled fine aggregate

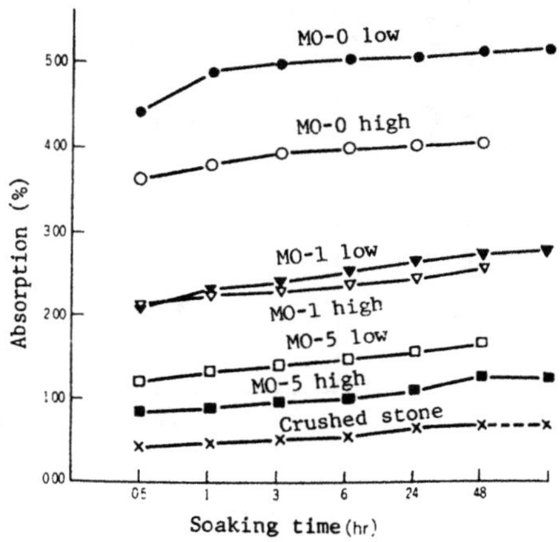

Figure 5 Soaking time and absorption of recycled coarse aggregate

3. The properties of concrete made from recycled aggregate

The fine aggregate percentage (s/a) in mix design for the recycled aggregate concrete could be fixed in a similar manner for concrete using original aggregates.

The dosage of AE admixture was not affected by coarse aggregate but it decreased when recycled fine aggregate was used in comparison to original fine aggregate was used. This appears to be caused by that the hardened paste adhering to the fine aggregate surface can very easily take air in the concrete.

The water content required to obtain a certain consistency was lower than the original aggregate was used. The cause is recycled coarse aggregate looses the angular shape of the original crushed stone by the process and grain shape is improved.

One example of the compressive strength of the recycled concrete is shown in Figure 6. The concretes made with the recycled fine/coarse aggregate show a lower strength in comparison with that of original fine/coarse aggregate when the cement/water ratio is same, and, in addition, a lower rate of strength increase caused by the increasing cement/water ratio. This tendency becomes more remarkable as the amount of sticking cement paste increases.

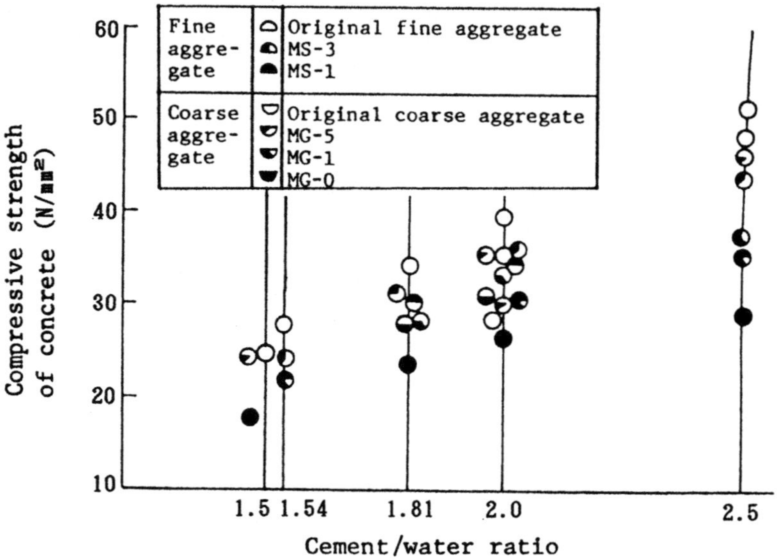

Figure 6 Cement/water ratio and compressive strength recycled aggregate

Table 4 shows the strength ratios for W/C 50 and 55%. This table suggests that the recycled aggregate used only for coarse aggregate results in a 15% strength reduction for MG-5 and 20% for MG-0. When the recycled aggregate is used only for fine aggregate, the recycled concrete seems to be subjected to a greater strength reduction.

Table 4 Ratio of compressive strength of concrete (%)

Fine aggregate	Coarse aggregate			
	Original aggregate	MG-5	MG-1	MG-0
Original aggregate	100	85	82	80
MS-3	91	91	75	-
MS-1	88	-	77	72

(Notes) 1) Figures for W/C=50 and 55%.

The results of freezing & thawing tests (ASTM C 666 A method) are given in Figure 7. In the case when the original aggregate was used for both the fine and coarse aggregates, the concrete shows a good performance with a D.F. of 96% after 300 cycles. When the recycled aggregate is used only for fine aggregate, the concrete seems to cause almost no trouble. When the recycled aggregate was used only for coarse aggregate, D.F. after 300 cycles is approximately 70% of the original aggregate concrete, and when it was used for both fine and coarse aggregates, the D.F. becomes approximately 50% of the original aggregate concrete.

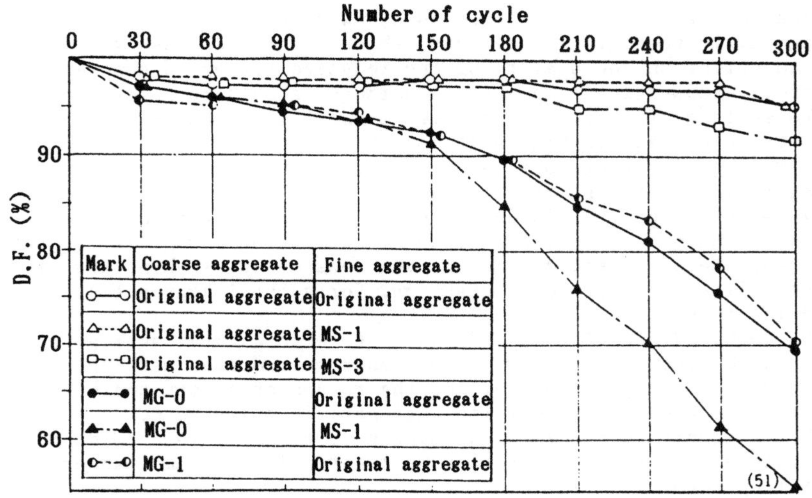

Figure 7 Result of freezing and thawing test

Furthermore, in this test, the use of the recycled aggregate resulted in higher scaling of the concrete surface. This scaling was especially marked when the recycled aggregate was used for both the fine and coarse aggregate. This cause seems to be that the low strength paste of the original concrete which was stuck on the aggregate brokeaway.

4. Guide for use of recycled aggregate concrete

Based on various laboratory tests and construction tests, we have prepared the "Guide for the Design and Use of Recycled Aggregate Concrete (draft)". In thise section, the concept of this guide will be introduced.

A higher degree of refining for the recycled aggregate can produce higher quality concrete, but this requires a higher manufacturing cost and a lower economical efficiency. For this reason the applicable concrete types are defined according to the degree of refinement of the recycled aggregates. As shown in Table 5 three coarse and two fine recycled aggregates are categorized according to absorption and stability in order to use them for three different types of concrete, which are shown in Table 7. The upper limits of the absorption and the lost mass are mainly defined on the basis of the concrete strength and durability.

With respect to the alkali-aggregate prevention, as the test method suitable for recycled aggregate is not specified yet, the measure such as limitation of total amount of Alkali in concrete which is used when reactive aggregate is used should be adopted.

In addition, items concerning the content of harmful substances in aggregates are set. When asphalt and brick are mixed with an original concrete to be utilized as a raw material of recycled aggregate, if they are not removed, these substances must give an adverse effect on the quality of the recycled aggregate concrete. Therefore, impurities like these substances should be removed.

The surface of a concrete structure built in a coastal area may be exposed to high salt concentration due to salt from the sea. This guide, therefore, does not allow the use of recycled fine aggregate in reinforced concrete.

Since fine powder produced during the recycled aggregate manufacturing process gives a minor effect on the durability of the concrete, the permissible limit of the loss due to washing test is set at a little higher level than that for general aggregates, as shown in Table 6.

The combined use of recycled aggregate with natural aggregate seems to improve the quality of the concrete compared with the single use of recycled aggregate, but for the reason of safety the mixed aggregate is treated as recycled aggregate in every case.

When a structure is designed using concrete made in accordance with this guide, the concrete can be treated in the same manner as natural aggregate.

Table 5 Qualities of recycled aggregates

By type	Recycled coarse aggregate				Recycled fine aggregate	
Item	1G	2G		3G	1G	2G
Absorption (%)	Less than 3	Less than 3	Less than 5	Less than 7	Less than 5	Less than 10
Lost mass	Less than 12	Less than 30	Less than 12 Note) (Less than 40)	-	Less than 10	-

(Note) In case that freezing & thawing durability is needless.

Table 6 Contents of harmful materials (weight percentage)

Loss in washing test	Recycled coarse aggregates 1G, 2G and 3G	Recycled fine aggregates 1G and 2G
Concrete surface is exposed to abrasion	1.5	5
Other cases	1.5	7

Table 7 Types of recycled concrete

Types of recycled concrete	Specified strength od recycled concrete	Coarse aggregate	Fine aggregate
I	More than 21 (N/mm^2) (reinforced concrete)	1G	Ordinary aggregate
II	More than 16 (N/mm^2) (plain concrete)	2G	Ordinary or 1G
III	Less than 16 (N/mm^2) (concrete subslab)	3G	2G

References

Hata, M., (1986) Improvement of Waste Concrete Recycling Technology. Civil Engineering Works

EXPERIMENTAL STUDIES ON PLACEMENT OF RECYCLED AGGREGATE CONCRETE

N. KASHINO and Y. TAKAHASHI
Materials Department, Building Research Institute, Ministry of Construction

Abstract
Various endeavors have been made to obtain recycled aggregates for concrete crushing concrete demolition wastes and rejected concrete. However, studies had not been made in the past regarding mix proportions and workabilities of concretes using recycled aggregates or the strength characteristics of such concretes when used in structures. The authors performed experiments on the placeabilities of recycled aggregate concretes using actual ready-mixed plants and structures, and made studies on adjustments of mixing water contents and air contents, pumpabilities, etc. As a result, it was shown that recycled aggregate concrete can be designed, manufactured, placed, and cured by the same methods as ordinary concrete.
Key words: Concrete waste, Recycled aggregate, Recycled aggregate concrete, Placement experiment, Actual structure, Placeability.

1. Introduction

Attempts to utilize concrete demolition wastes for recycled coarse aggregate for concrete (hereinafter called recycled aggregate) have been made from a long time ago, and many experimental studies have been conducted to investigate the properties of recycled aggregate concrete, but experiments on manufacture of ready-mixed concrete and placement in actual structures have not been carried out up to now. The authors, based on the results of tests performed in the laboratory on the physical properties of recycled aggregate concretes, conducted experiments to ascertain whether recycled aggregates are indeed usable by actually manufacturing ready-mixed concrete and placing it in actual structures, investigating placeability and the strength characteristics when made into a structure, and these are reported below.

2. Outline of laboratory experiments

An outline of the results of laboratory experiments on recycled aggregate concrete conducted by Kashino et al. (1985,1986) is given below.

2.1 Physical properties of recycled aggregate

The specific gravities, absorptions, and unit weights of recycled aggregates made by crushing concrete wastes and adjusting to fineness modulus of around 6.5 are given in Table 1.

Table 1. **Specific gravities, water absorptions, unit weights of recycled aggregates (average values given in the table)**

Crushing	Specific gravity	Absorption (%)	Unit weight (kg/ℓ)
Jaw crusher	2.38	5.98	1.36
Jaw crusher and impact crusher combined	2.42	5.47	1.34
JIS on crushed stone for concrete	not less than 2.5	not more than 3	-

With the recycled aggregates there was adherence of cement paste on the original aggregate of approximately 21 percent in case of large particles and approximately 28 percent in case of small particles. Specific gravities were slightly lower than for ordinary crushed stone, and absorption rates **relatively high.** According to various tests performed up to this time, there are hardly any differences due to water-cement ratios of the original concretes or the methods of crushing. Crushed stone and recycled aggregates are shown in Photo. 1.

Photo. 1. Left—recycled aggregate. Right—crushed stone.

There are cases when plaster and/or paint are adhered to the recycled aggregate. Concrete strengths were lower by 10 to 20 kgf/cm^2 in terms of compressive strength in case of 15 to 20 percent adherence of foreign matter. Recycled aggregates free of foreign matter were used in the placement tests reported here.

2.2 Physical properties of recycled aggregate concrete

The mix proportions of concrete were made the same throughout for air-entrained concrete of water-cement ratio 0.60 and slump 18 cm except that the ratios of recycled aggregate used were varied to see how physical properties would be affected. It was found that the same workability is obtained with unit water content roughly the same as for ordinary cases, that bleeding is smaller as the ratio of recycled aggregate becomes higher, and that, as shown in Fig. 1, when the ratio of recycled aggregate mixed in coarse aggregate is lower than 30 percent, strength characteristics such as compressive strength, modulus of elasticity and creep, and resistance to frost damage are practically unchanged compared with ordinary concrete.

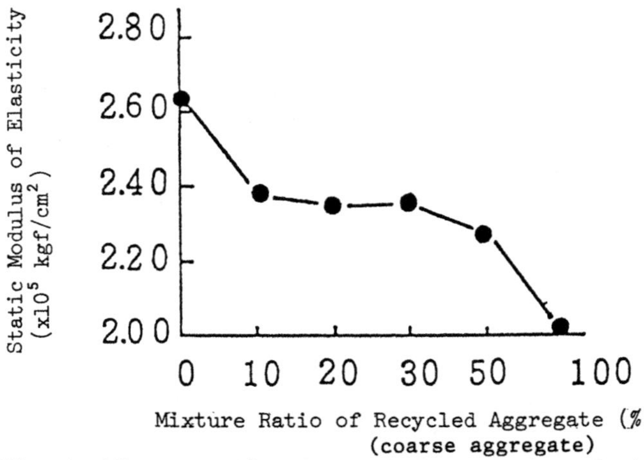

Fig. 1 Mixture ratio of recycled aggregate and strength characteristics of concrete (52 wks).

When the ratio of recycled aggregate mixture becomes large, the amount of drying shrinkage becomes about 10 percent greater than for ordinary concrete. However, permeability and the rate of carbonation are almost unchanged from those of ordinary concrete even when the mixture ratio is high.

3. Placement experiments using actual structures

3.1 Outline of experiments

Keeping the matters mentioned in Section 2 in mind, placement experiments were conducted with actual structures using ordinary concrete containing crushed stone as coarse aggregate, concrete with 30 percent recycled coarse aggregate, and concrete with 100 percent recycled coarse aggregate. The prototype concrete was of the mix proportions shown in Table 2.

Table 2. Outline of prototype recycled aggregate concrete used in placement experiments.

Estimated mix proportions	Cement	332 kg/m^3
	Coarse aggregate	1,200 "
	Fine aggregate	507 "
Compressive strength of core (ϕ10 x 20 mm) (average)		287 kgf/cm^2

Ordinary portland cement was used as the cement, river sand as the fine aggregate, and the mix proportions shown in Table 3 were selected based on trial mixes. Slump was made 18 cm and air content 4 percent. In case of using recycled aggregate, an air-entraining aid at a rate of 0.001 percent of cement was additionally required to adjust air content. Prewetting and adjustment of water content were also needed in case of recycled aggregate.

Table 3. Mix proportions of ready-mixed concrete.

Variety of concrete	W/C	s/a (%)	Water (ℓ/m^3)	Cement (ℓ/m^3)	Sand (ℓ/m^3)	Crushed stone (ℓ/m^3)	Recycled agg. (ℓ/m^3)	AEA (Cx%)
Ordinary concrete	0.60	45	172	91	309	378	-	0.2
30% recycled aggregate concrete	0.60	45	172	91	309	265	113	0.2
100% recycled aggregate concrete	0.60	45	172	91	289	-	398	0.2

The concretes were manufactured at a ready-mixed plant and placement was done by pump in two structures 3 m in frontage, 4 m in depth, and 3 m in height, and three walls 2 m by 1.5 m, 0.15 m in thickness (see Photo. 2). Consolidation was done by form vibrator operated for 10 sec to apply equal energies to the concrete.

Photo. 2. Actual structures in placement experiments of recycled aggregate concretes.

The items of test were slump, air content, unit weight, compressive strength after one year, static modulus of elasticity, and pumpability and form fillability.

3.2 Results of tests

The results of tests on slumps, air contents, and unit weights of ready-mixed concrete when unloaded and one hour later on finishing placement are shown in Table 4.

Table 4. Results of tests on ready-mixed concrete.

Variety of concrete	When tested	Slump (cm)	Air (%)	Unit weight (kg/ℓ)
Ordinary concrete	Unloaded	17.5	4.8	2.34
	Placed	16.5	4.6	-
30% recycled aggregate concrete	Unloaded	18.0	4.5	2.31
	Placed	19.5	4.5	-
100% recycled aggregate concrete	Unloaded	20.5	4.8	2.23
	Placed	21.5	4.6	-

The compressive strength and static modulus of elasticity test results of concretes in the actual structures after one year are shown in Table 5, while the strength characteristics at one year of specimens made separately with the same concretes are shown in Table 6.

Table 5. Results of strength characteristics tests with core specimens from actual structures. (Average values of 3 specimens)

Variety of concrete	Cored location		Compressive strength (kgf/cm^2)	Static modulus of elasticity ($x10^5$ kgf/cm^2)
Ordinary concrete	Structure	Upper	312	1.96
		Lower	328	2.03
	Wall	Upper	345	2.20
		Lower	349	2.17
30% recycled aggregate concrete	Structure	Upper	306	2.14
		Lower	364	2.28
	Wall	Upper	321	2.25
		Lower	380	2.39
100% recycled aggregate concrete	Structure	Upper	280	2.17
		Lower	318	2.50
	Wall	Upper	309	2.32
		Lower	350	2.30

3.3 Considerations

It was learned through the series of experiments described in the foregoing that ready-mixed concrete using recycled aggregate can be designed and manufactured by ordinary methods. However, in case of recycled aggregate concrete, it is necessary to add a small amount of air-entraining aid to adjust air content.

As shown in Table 4, slump loss and air content loss did not occur at all in recycled aggregate concrete during the one hour from the time of unloading ready-mixed concrete until completion of placement. It was the same in tests of 100-percent recycled aggregate concrete performed separately.

Meanwhile, as Table 5 shows, it was learned that the compressive strength of concrete containing 30 percent of recycled aggregate is approximately the same as with ordinary concrete. This was the same as the results of laboratory experiments. There were also correlations between the strength characteristics of core specimens taken

Table 6, Results of strength characteristics tests with standard specimens. (Averages of 3 cylinder specimens field-cured with curing compound)

Variety of concrete	Compressive strength (kgf/cm^2)	Static modulus of elasticity ($x10^5$ kgf/cm^2)
Ordinary concrete	325	2.58
30% recycled aggregate concrete	359	2.68
100% recycled aggregate concrete	358	2.68

from actual structures and the strength characteristics of standard specimens given in Table 6. Although slight differences were seen in concrete strengths between upper and lower parts of structures, this may be considered natural when settlement of concrete is taken into account. Hardly any difference could be seen between the two in the degrees of settlement.

The pumpability of recycled aggregate concrete was more or less the same as that of ordinary concrete, and trouble such as clogging of the pipeline during pumping was not seen. The effects of consolidation using vibrators did not differ either so far as seen in visual inspection. Defective parts such as honeycombing did not exist in actual structures and walls. The same can be said of concretes using 100 percent recycled aggregates tested separately.

4. Conclusions

The following were disclosed through the placement experiments using recycled aggregates just conducted.

1. Regarding mix proportioning of concrete, recycled aggregate can be handled in approximately the same manner as ordinary crushed stone.
2. Handling of recycled aggregate concrete at the ready-mixed plant can be about the same as with ordinary concrete.
3. The placeability of recycled aggregate concrete is not different from that of ordinary concrete.
4. The strength characteristics obtained in laboratory experiments are reflected without change in actual structures.

In essence, it was found that with recycled aggregate concrete, there would be no problems in particular concerning placeability and strength characteristics when manufacturing and placing the concrete by methods the same as ordinary ones. There are, however, problems which remain to be studied regarding stockyards, solidification during stocking and the like.

The authors are indebted to Messrs. Motoo Hisaka and Kei Yanagi of the Japan Testing Center for Construction Materials, and Mr. Muneo Nakagawa of Toda Construction Co., Ltd. for their cooperation in carrying out the experiments, and many thanks are hereby extended.

References

Kashino,N.,Tanaka,H.,Hisaka,M.,and Yanagi,K.(1985) Experimental study on physical properties of concrete using recycled aggregate, Summaries of Technical Papers of Annual Meeting,Architectural Institute of Japan.

Building Research Institute,Ministry of Construction (1986) Report on studies concerning recycling technologies for waste materials in construction project.

STRENGTH AND ELASTIC MODULUS OF RECYCLED AGGREGATE CONCRETE

Dr. M. Kakizaki — Kajima Corp., Japan
M. Harada — Kajima Corp., Japan
Dr. T. Soshiroda — Shibaura Institute of Technology, Japan
S. Kubota — Ohbayashi Corp., Japan
T. Ikeda — Takenaka Komuten Co., Ltd., Japan
Dr. Y. Kasai — Nippon University, Japan

Abstract
This study was conducted to clarify compressive strength, elastic modulus and bonding strength of steel bars for recycled aggregate concrete. This aggregate was produced by crushing a particular type of concrete into the specified particle size using a jaw crusher, or by some other method. The results were compared with comparable figures for normal weight concrete (NWC), and the possibility of applying concrete made from recycled aggregate to structures was discussed. The following results were obtained from the tests: a) Compressive strength of recycled aggregate concrete is lower than that of NWC by 14 to 32%; b) The relationship between compressive strength and cement-water ratio is linear; c) Elastic modulus of recycled aggregate concrete is lower than that of NWC by 25 to 40%; and d) Bonding strength of vertical steel bars is between 2.4 and 3.7 times that of horizontal steel bars.

Key words: Recycled aggregate, Recycled aggregate concrete, Strength, Elastic modulus, Bonding strength

1. Introduction

Recently, the number of concrete structures demolished has been increasing rapidly, and concrete disposal is a problem because the conventional method of using it for land reclamation is becoming more difficult. Therefore, recycled aggregate concrete is important in terms of economy and the efficient use of resources because concrete waste, which is generated in plenty, can be reused in new structures after being crushed.

In this study, tests were performed on recycled aggregate concrete concerning compressive strength, elastic modulus and bonding strength of steel bars. The results were compared with corresponding figures for NWC.

2. Test Plan

2.1 Method

Table 1 shows the factors of the tests and characteristic values.

Table 1 Factors of Tests and Characteristic Values

Factors	Level						Characteristic values	Constant conditions
	1		2		3			
Water-to-cement ratio of recycled concrete aggregate (%)	45		55		68		Compressive Strength Elastic Modulus Bonding strength	Slump (8, 15, 21 cm) Air content (4%) Curing condition (wet air-dry)
Fine aggregate, Coarse aggregate	Natural aggregate		Recycled concrete aggregate		-			
Water-to-cement ratio of mix (%)	45	50	55	60	65	70 (75)		
Age (weeks)	1		4		13			

2.2 Materials

(a) Cement: Ordinary Portland cement (See Table 2.)

Table 2 Physical Properties of Ordinary Portland Cement

Specific gravity	Blaine specific surface (cm^2/g)	Setting Time: Water content (%)	Setting Time: Initial (hr.-min)	Setting Time: Final (hr.-min)	Flow (mm)	Compressive strength (Kgf/cm^2): 3 days	7 days	28 days
3.15	3160	27.5	2-21	3-39	237	130	220	402

(b) Aggregate: The types of aggregate are shown in Table 3. Recycled concrete aggregate was manufactured according to the dimensions 2505 of crushed stone prescribed by JIS A 5005 (crushed stone for concrete) by crushing with a jaw crusher of pitch 3.3 mm. The crushed stone was divided into fine aggregate and coarse aggregate. Physical properties of the aggregate are shown in Table 4.

Table 3 Types of Aggregate

Material	W/C ratio of recycled concrete aggregate (%)	Fine aggregate	Coarse aggregate
Natural aggregate	-	NS	NG
Recycled concrete aggregate	45	CS45	CG45
	55	CS55	CG55
	68	CS68	CG68

Table 4 Physical Properties of Aggregate

Material	Type	Specific gravity: Surface dry	Specific gravity: Dry	Water absorption (%) 1)	Bulk density (Kg/l) 2)	Solid volume (%) 3)	Fineness modulus 4)
Fine aggregate	NS	2.62	2.58	3.31	1.77	67.5	2.91
	CS45	2.25	2.03	11.5	1.29	57.1	3.83
	CS55	2.29	2.08	10.9	1.32	57.5	3.72
	CS68	2.27	2.06	11.5	1.30	57.3	3.73
Coarse aggregate	NG	2.67	2.65	1.37	1.74	65.1	6.90
	CG45	2.46	2.31	5.82	1.35	55.8	6.88
	CG55	2.45	2.30	5.81	1.36	56.6	6.84
	CG58	2.45	2.32	5.85	1.37	56.7	6.84

Note Method of test:

1) JIS A1109 (Fine Aggregate)
JIS A1110 (Coarse Aggregate)

2),3) JIS A1104 4) JIS A1102

(c) Admixture: AE agent (Vinsol)

2.3 Mix Proportion

The mix proportion for recycled aggregate cncrete was determined through trial mixing to check workability according to the mix proportion table of JASS 5 crushed stone-AE concrete (1973 version). This defined the percentage of fine aggregate. Table 5 shows the types of concrete and mix proportions.

Table 5 Types of Concrete and Mix Proportions

Types of concrete: Fine aggregate	Types of concrete: Coarse aggregate	Slump (cm)	Water-to-cement ratio (%): 45	50	55	60	65	70	75
NS	NG		O	O	O	O	O	O	O
NS	CG45		O		O	O	O		
NS	CG55	8	O		O	O	O		
NS	CG68		O		O	O	O	O	
NS+CS45	CG45		O			O			
NS+CS55	CG55	15		O		O			O
NS+CS68	CG68					O		O	
CS45	CG45	21			O	O	O		
CS55	CG55		O			O	O		
CS68	CG68				O	O	O		

2.4 Test Method

Table 6 and Fig. 1 show the test method.

Table 6 Specifications of Concrete Specimens and Method of Testing

Test item	Specifications of concrete specimens and method of testing	Dimensions of specimens
Compressive strength	JIS A1108	10φ x 20cm
Elastic modulus	To calculate elastic modulus, the vertical strain of compressive strength was measured with a compressometer, and 1/3 of the maximum value was used.	10φ x 20cm
Bonding strength	1) Specimens were produced as shown in Fig.1 2) Round steel bars (19φ) were used. 3) The test method was based on the standard ASTM-C234-57T.	Vertical steel bar tests: 15cm cubic Horizontal steel bar tests: 15x15x45cm

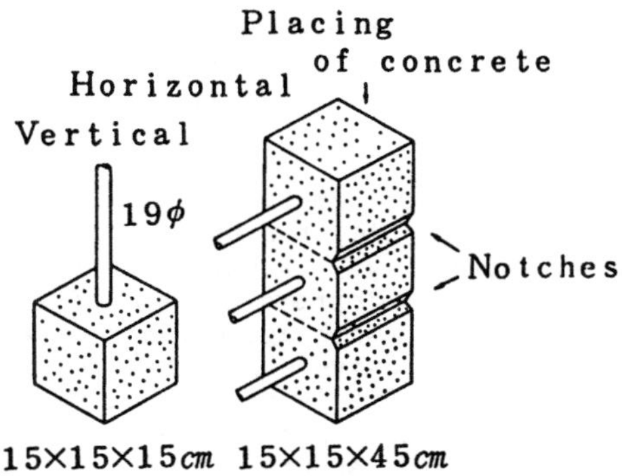

Fig.1 Specimens for Bonding Test (pulling type)

3. Test Results and Appraisal

3.1 Compressive Strength

1) Relationship between Compressive Strength and Age

Fig. 2 shows compressive strength as a percentage of that at age 4 weeks. Compared with compressive strength of NWC (NS·NG), the compressive strength of recycled aggregate concrete at age 4 weeks is about 14% lower for NS·CG concrete, about 25% lower for (NS +CS)·CG concrete and about 32% lower for CS·CG concrete. It is considered that the lower strength of recycled aggregate concrete is due to faults, including many microcracks, in the mortar of the recycled concrete aggregate. A similar tendency can be noted with lightweight aggregate concrete, and it is necessary to sidestep this problem by selecting the method of usage and application according to aggregate quality.

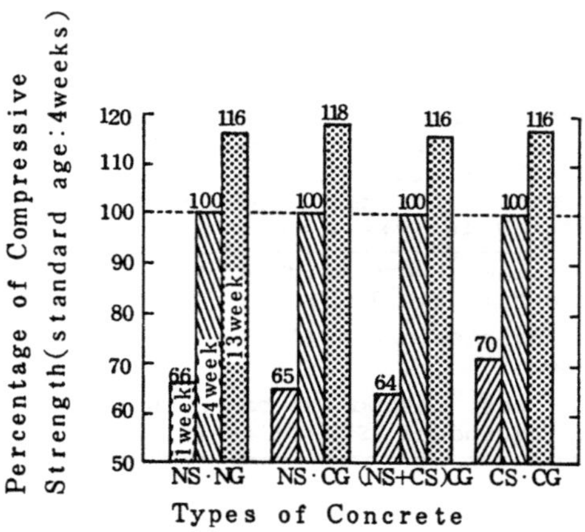

Fig.2 Relationship between Percentage of Compressive Strength at Age 4 Weeks and Type of Concrete (W/C: 45~70%)

Moreover, when the development of compressive strength with age is compared for various types of concrete, the strength of the recycled aggregate concrete is about 65% at age 1 week, about 116% at age 13 weeks, showing similar strength development to NWC. In addition, development of compressive strength with age differs by water-to-cement ratio: at lower ages, compressive strength increases with decrease in water-to-cement ratio. It is considered that this is due to the effect of hardening speed and volume of cement paste.

2) Relationship between Water-to-Cement Ratio (Cement-Water Ratio) and Compressive Strength

Fig. 3 shows the relationship between cement-water ratio and compressive strength for various types of concrete. The results show that compressive strength increases linearly with cement-water ratio. However, compressive strength for recycled aggregate concrete (NS·CG, (NS+CS)·CG, CS·CG) did not show as much increase as for NWC. For ordinary water-to-cement ratio of 60%, compressive strength was 240 to 280 kgf/cm^2 for NS·CG concrete, about 220 kgf/cm^2 for (NS+CS)·CG concrete, and 170 to 220 kgf/cm^2 for CS·CG concrete. Moreover, the compressive strength of recycled aggregate concrete is a smaller percentage of that of NWC at higher cement-to-water ratios. As mentioned above, it is considered that the bonding strength between recycled concrete aggregate and cement paste, as well as cracks, etc. in the cement paste which are incurred during crushing have an effect.

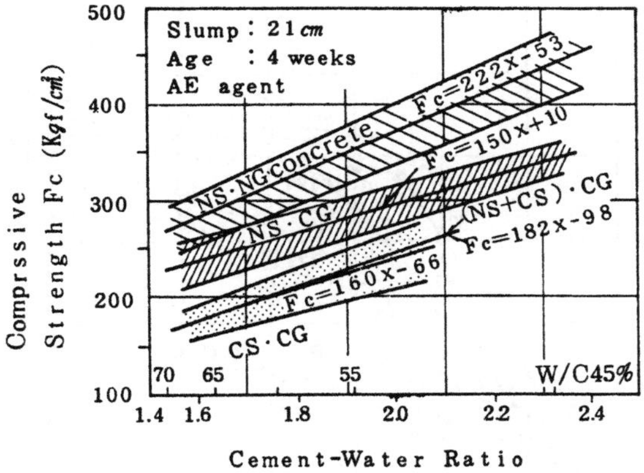

Fig.3 Relationship between Cement-Water Ratio and Compressive Strength

The relationship between cement-water ratio and compressive strength can be expressed by the following linear equations.

Normal weight concrete	(NS·NG)	$F_c = 222X - 53$	(1)
Recycled aggregate concrete	(NS·CG)	$F_c = 150X + 10$	(2)
	((NS+CS)·CG)	$F_c = 182X - 98$	(3)
	(CS·CG)	$F_c = 160X - 66$	(4)

Where F_c: Compressive strength (kgf/cm^2) of concrete at age 4 weeks

X: Cement-water ratio

3) Relationship between Compressive Strength and Method and Period of Curing of Specimens

Fig. 4 shows the relationships between percentage of compressive strength of recycled aggregate concrete (CS·CG) at age 4 weeks after wet curing and age for different curing methods and testing conditions. (Figures for NS·NG and NS·CG concrete have been omitted due to lack of space.) The results show that the compressive strength of concrete increases if wet curing is used over a long period. Compressive strength decreases temporarily but then increases with age if the concrete is cured in air-dry conditions at some age. For standard curing, strength at age 13 weeks is 108 to 113% of that at age 4 weeks. For air drying, the figures are about 125%. In addition, if concrete being wet cured is dried immediately before testing, compressive strength at lower ages is larger, while there is no significant difference at age 13 weeks. Comparing compressive strengths at age 4 weeks in this case, strength in wet conditions is about 8% larger than in dry conditions.

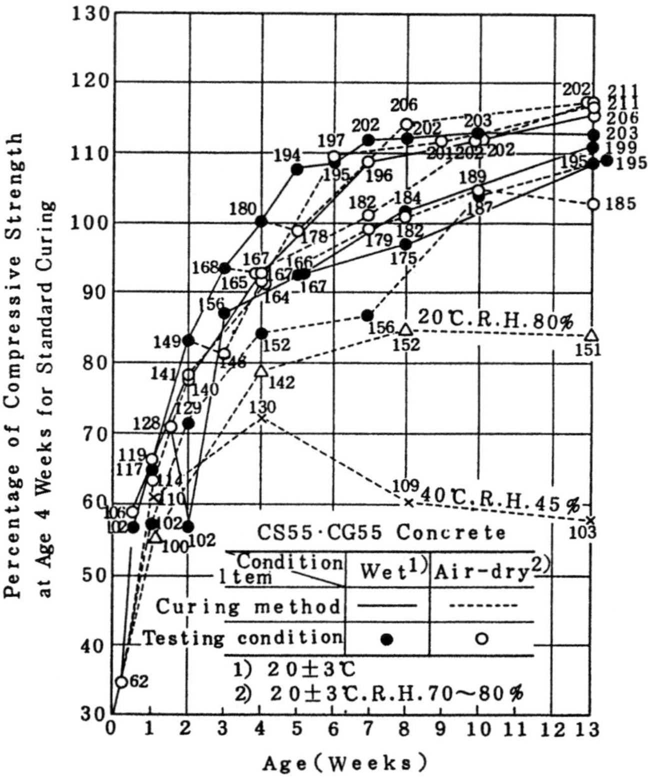

Fig.4 Relationship between Percentage of Compressive Strength at Age 4 Weeks for Standard Curing and Age (CS55·CG55 Concrete)

3.2 Relationship between Compressive Strength and Elastic Modulus

Figs. 5 shows the relationship between compressive strength and elastic modulus (E1/3). In the figure, the solid lines for specific gravity γ =1.9 to 2.3 represent elastic modulus obtained by the structural formula of the Architectural Institute of Japan (AIJ) with variables concrete specific gravity (γ) and strength (f_c). The formula given by the Institute is for concrete specimens with curing by air drying, while the experimental results of the tests in this study are from specimens with a dry surface but saturated with water inside. According to the test results, the elastic modulus of NWC is between 2.5 and 3.9 x 10^{-4} kgf/cm^2, which is larger than the value estimated from the formula given by the Institute. On the other hand, the elastic modulus of recycled aggregate concrete is between 1.4 and 3 x 10^{-4} kgf/cm^2. This is within the range specified by γ = 1.9 to 2.3 using the Institute's formula, and the calculated value and the measured value were almost the same. In addition, Fig. 6 shows that, compared with the value at age 4 weeks, elastic modulus at age 1 week is about 12% lower for NS·NG concrete, about 15% lower for NS·CG concrete, about 19% lower for (NS+CS) CG concrete and about 8% lower for CS·CG concrete. Accordingly, elastic modulus for recycled aggregate concrete shows almost the same tendencyas for lightweight aggregate concrete. This can be explained by the state of bonding between the recycled concrete aggregate and cement paste, and cracks in the cement paste.

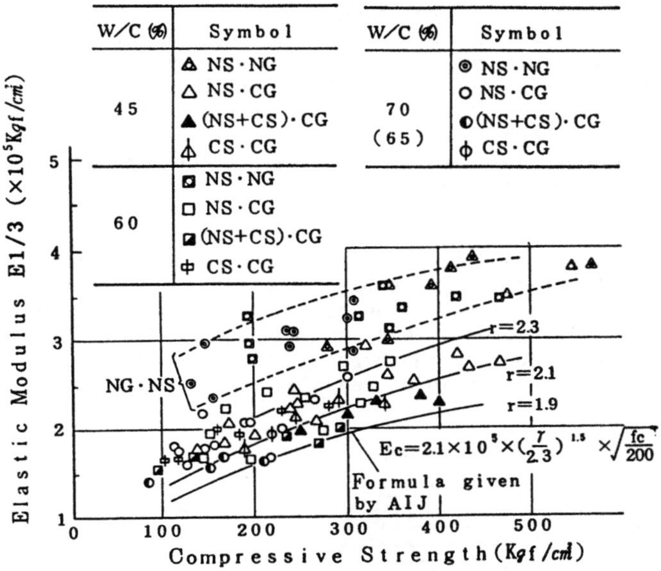

Fig.5 Relationship between Compressive Strength and Elastic Modulus(E1/3)

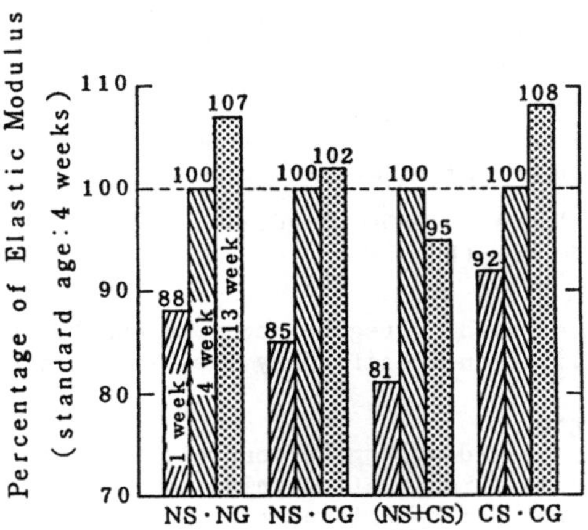

Fig.6 Relationship between Percentage of Elastic Modulus at Age 4 Weeks and Type of Concrete (W/C: 45~70%)

3.3 Bonding Strength of Steel Bars

Fig. 7 shows bonding strength of recycled aggregate concrete as a percentage of that of NWC. The bonding strength of vertical steel bars of recycled aggregate concrete (CS·CG) is between 18.2 and 18.6 kgf/cm^2, which is about 25% lower than for NWC (24.2 kgf/cm^2). The values are lower in about the same degree as compressive strength. Moreover, the bonding strength of horizontal steel bars is between 5 and 7.5 kgf/cm^2, which is about 3 to 35% lower than for NWC (7.66 kgf/cm^2). Generally, the decrease in percentage of bonding strength for horizontal steel bars is smaller than for vertical bars. The difference can be attributed to the following. The bonding strength of vertical steel bars is greatly affected by friction with the concrete. Recycled aggregate concrete has more cement paste and so it is not as hard as NWC, resulting in less friction, and a decrease in bonding strength. Also, the bonding strength of horizontal steel bars is greatly affected by the settling property of fresh concrete. Therefore, with horizontal bars, the bonding strength for recycled aggregate concrete is approximately the same as for NWC since there is little bleeding. However, the bonding strength does not vary with change in water-to-cement ratio of recycled concrete aggregate.

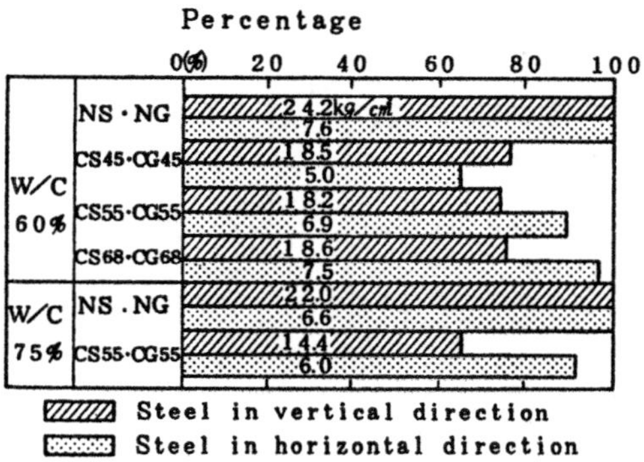

Fig.7 Percentage of Bonding Strength

4. Conclusion

It was clarified that recycled aggregate concrete can be used within a certain range of application specified by the limitations below.

4.1 Compressive Strength

1) The strength of recycled aggregate concrete does not change greatly with water-to-cement ratio of recycled concrete aggregate.

2) The relationship between cement-water ratio and compressive strength is linear, as shown in Eqs. (1) to (4).

3) The upper limit of compressive strength at water-to-cement ratio 45% is 350 kgf/cm^2 for NS·CG concrete, 300 kgf/cm^2 for (NS+CS)·)CG concrete and 280 kgf/cm^2 for CS·CG concrete.

4) The strength at age 4 weeks of recycled aggregate concrete for

the range of water-to-cement ratio between 40% and 75% is lower than that of NWC by about 14% for·NS CG concrete, by about 25% for (NS+CS)·CG concrete and by about 32% for CS·CG concrete. This should be considered in case of actual applications.

5) Compressive strength increases if the concrete is cured in wet conditions over a long period of time. Compressive strength decreases temporarily but then increases with age if the concrete is cured in air-dry conditions at some age.

4.2 Elastic Modulus

1) Elastic modulus for recycled aggregate concrete is lower than that of NWC by about 25 to 40%.

2) Elastic modulus (E1/3) at age 1 week is lower than that at age 4 weeks by about 12% for NS·NG concrete, by about 15% for NS·CG concrete, by about 19% for (NS+CS)·CG concrete and by about 8% for CS·CG concrete. This tendency is similar to that of lightweight aggregate concrete.

3) The relationship between compressive strength and elastic modulus can be expressed by the following equation given by the Architectural Institute of Japan.

$$E_c \doteqdot 2.1 \times 10^5 \times (\gamma/2.3)^{1.5} \times \sqrt{f_c/200}$$

4.3 Bonding Strength of Steel Bars

The bonding strength for recycled aggregate concrete is lower than for NWC by about 25% for vertical steel bars and by 3 to 35% for horizontal steel bars, and the decrease in percentage of bonding strength is smaller for horizontal bars than for vertical bars.

Acknowledgements

This study was undertaken by the Committee for Treatment and Reuse of Construction Waste, the Building Constructors Society. The authors would like to express their sincere gratitude to everyone who cooperated or assisted with the study.

References

1) Dr. M. Kakizaki, Dr. T. Soshiroda, H. Harada, M. Ikeda and A. Sakamoto, Oct. 1975, "Study of Concrete Using Crushed Concrete Aggregate (Part 4)," Proceedings, Lecture Meeting of Architectural Institute of Japan, pp.331 - 332
2) S. Kubota, Dr. Y. Kasai, T. Soshiroda, Dr. M. Kakizaki and M. Ikeda, Oct. 1976, "Study of Concrete Using Crushed Concrete Aggregate (Part 9), Proceedings, Lecture Meeting of Architectural Institute of Japan, pp.55 - 56
3) Dr. Y. Kasai, Feb. 1976, "Reuse of Crushed Concrete (Part 2: Concrete using Crushed Aggregate)", Cement & Concrete, No.348, pp. 16 - 28

METHODS OF IMPROVING THE QUALITY OF RECYCLED AGGREGATE CONCRETE

R. SRI RAVINDRARAJAH and C.T. TAM
Department of Civil Engineering, National University of Singapore

Abstract
Waste concrete with a lower degree of contamination is a potential source for the production of aggregate for concrete. Recycled aggregate particles consist of substantial amount of relatively soft cement paste component. These aggregates are more porous and less resistant to mechanical actions. The limitation of using recycled aggregates in new concrete is their detrimental influence on the concrete properties. In comparison with natural aggregate concrete recycled aggregate concrete shows reductions in strength and modulus of elasticity and increases in drying shrinkage and creep. Recycled aggregate concrete may also be less durable due to increase in porosity and permeability. Economical ways of improving the quality of recycled aggregate concrete are: (i) by reducing the water-cement ratio; (ii) by reducing the water content using a water reducing admixture without affecting the workability; (iii) addition of pozzolan; and (iv) blending of recycled aggregates with the natural aggregates.
Key words: Recycling, Strength, Modulus of elasticity, Fly ash, Silica fume, Mix composition, Drying shrinkage, Aggregate

1. Introduction

Recycling is a process of recovery and subsequent use of a material for the manufacture and/or fabrication of the same or similar product from which the waste was originated. Recycling of materials used in outdated construction is probably as old as civilization itself.

In many countries, a considerable amount of demolition waste is generated and concrete forms a significant proportion of the waste. The 'waste concrete' is also generated from many other sources. In the precast concrete industry where wastage may arise from breakage of precast elements during production and transporting. In the central and site testing laboratories contamination free crushed concrete are obtained from the control test specimens. At construction sites waste concrete is generated from demolition of rejected construction due to poor quality workmanship or unauthorized work. Waste concrete, when adequately reduced in size, can be used as aggregates for concrete production.

Crushed concrete coarse particles can be used to replace the good quality natural aggregate in new concrete. It is also possible to replace the natural sand with crushed concrete fines. Hansen (1986) reviewed the published information on the properties of recycled aggregate and recycled aggregate concrete.

The necessity for the use of recycled aggregate in concrete arises due to the following reasons: (a) diminishing supplies of natural aggregates; (b) securing ample supply of concrete aggregates to the construction industry; (c) diminishing dumping area within the urban limits; and (d) avoiding danger to marine life by limiting the indiscriminate dumping of highly alkaline nature of concrete in the sea.

2. Recycled aggregate

2.1 Components of recycled aggregate

In general, the particles in the concrete crusher product consist of natural aggregate particles partially or fully coated with cement paste. BCSJ (1978) reported that the amount of cement paste attached to the original aggregate decreases with the increase in the size of the recycled aggregate. About 20% of cement paste is attached to recycled aggregate particles between 20 to 30 mm size, while the concrete fines below 0.30 mm size contain 45 to 65% of old cement paste.

Hansen and Narud (1983a) reported the volume percent of mortar attached to gravel particles in recycled aggregate to be between 25 and 35% for 16-32 mm size, and around 40% for 8-16 mm size, and around 60% for 4-8 mm size. Sri Ravindrarajah and Tam (1985) found the volume percent of mortar attached to granite particles in 5-20 mm size recycled aggregate to be around 50%. The higher proportion for granite particles than that for gravel particles may be due to the better aggregate-cement paste bond in the former.

Crushed concrete fines consist of crushed aggregate particles and partially hydrated cement paste which is made up of unhydrated cement grains and hydration products of cement. Sri Ravindrarajah and Tam (1987a) observed that when the concrete fines come in contact with water, the alkali content of the water begins to increase. This is due to the soluble calcium hydroxide in the concrete fines, and to the newly formed on hydration of unhydrated cement grains. Hansen and Narud (1983b, 1983c) reported that the ground crushed concrete fines can be used to produce bricks with the compressive strength of upto 25 MPa using autoclave method.

2.2 Physical properties of recycled aggregate

Since the recycled aggregate particles consist of relatively porous cement paste component, the specific gravity of these particles is lower than that for granite. The apparent specific gravity values on the oven-dried basis (105°C) are 2.67, 2.61, 2.49 and 2.32 for the granite, natural sand, recycled coarse aggregate, and concrete fines, respectively, as reported by Sri Ravindrarajah and Tam (1987b).

24 h absorption value for the recycled coarse aggregate which was obtained by crushing an year old concrete is 5.68%. About 80% of the 24 h absorption was observed to occur within the first 5 minutes for the oven-dried (105°C) aggregate, when soaked in water at the room

temperature of 30° C. Fig. 1 shows that the absorption capacity of recycled aggregate is increased with the decrease in the aggregate size. Quality of the waste concrete seems to have less influence on the absorption capacity of recycled aggregate. The results implies that the volume content of cement paste in the aggregate particles is increased when the aggregate size is reduced.

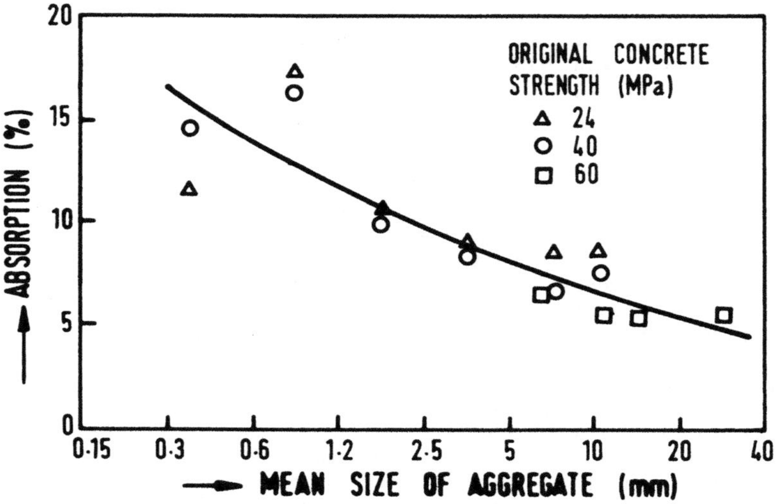

Fig. 1. Effect of aggregate size on absorption capacity for recycled aggregate

2.3 Mechanical properties of recycled aggregate

Table 1 summarizes the mechanical properties of recycled coarse aggregate which was obtained by crushing an year old concrete in a laboratory jaw crusher. Various tests for the mechanical properties were conducted in accordance with the relevant BS or ASTM methods.

Table 1: Mechanical properties of recycled coarse aggregate

	Granite	Recycled aggregate			
Property	19.0 to 5.0 mm	19.0 to 13.2 mm	13.2 to 9.5 mm	9.5 to 5.0 mm	below 5.0 mm
ASTM Abrasion value (%)	18.5	28.0	27.7	31.6	36.2
BS 10% Fine load (MPa)	18.2	11.0	11.0	10.0	8.0
BS Impact value (%)	15	27	25	29	28
Angularity No.	11	16	16	17	16

The recycled aggregate particles are more angular than the crushed granite particles, probably due to the ease of breaking in an irregular shape along the soft cement paste component in concrete. The higher values for the abrasion and impact resistance and lower value for 10% fine crushing load for recycled aggregate than those for granite indicate the weakness of the recycled aggregate against mechanical actions. The results also show that the mechanical resistance of recycled aggregate is decreased with the reduction in the maximum aggregate size. This is not surprising as the volume content of relatively soft cement paste component in the recycled aggregate particles is increased with the decrease in the aggregate size.

3. Effects of using recycled aggregate on concrete properties

Since 1983 extensive studies have been conducted at the National University of Singapore on the use of recycled aggregate in new concrete and the results have been reported elsewhere (Sri Ravindrarajah and Tam (1985, 1986, 1987a, 1987b, 1988); Sri Ravindrarajah (1986); Ong, Liaw and Tam (1985); Ong and Sri Ravindrarajah (1987)). The main conclusions of these studies are summarized below.

(a) For a constant degree of workability, the water requirement for recycled aggregate concrete is about 10% more than that for a similar natural aggregate concrete.

(b) Use of recycled aggregate instead of natural aggregate as coarse aggregate on concrete properties are: reduction in compressive strength upto 25%; reduction in modulus of elasticity upto 30%; considerable increases in drying shrinkage and creep; increase in damping capacity; and reduction in pulse velocity and fracture toughness.

(c) Use of crushed concrete fines instead of natural sand as fine aggregate on concrete properties are: marginal effect on compressive strength; upto 20% reduction in modulus of elasticity; upto 40% increase in drying shrinkage; and marginal increase in creep.

(d) Use of recycled coarse aggregate and crushed concrete fines instead of natural aggregates on concrete properties are: about 10% reduction in compressive and tensile strengths; upto 35% reduction in modulus of elasticity; and nearly 100% increase in drying shrinkage.

(e) Relationships between modulus of elasticity and compressive strength, and pulse velocity and compressive strength are affected when recycled aggregates are used instead of natural aggregates. For equal strength concretes, recycled aggregate concrete shows lower values for modulus of elasticity and pulse velocity than those for natural aggregate concrete.

(f) The detrimental effects of using crushed concrete fines in concrete can be mitigated by a partial replacement of the concrete fines with pulverised fuel ash.

4. Experimental studies

Experimental study on the recycled aggregate concrete properties was carried out in two series. In the first series, a medium strength concrete, having the cement content of 345 kg/m3 and the water-cement ratio of 0.57 by weight was chosen in this study. In the control concrete mix (Mix GS) crushed granite and natural sand were used as aggregates. Recycled concrete aggregates which were obtained from an year old contamination free concrete were used as coarse and fine aggregates in producing the recycled aggregate concrete mix (Mix RR). Two more mixes with recycled aggregates, having either 10% low calcium fly ash (Mix RRF) or 5% condensed silica fume (Mix RRS) additions, by cement weight, were also produced.

In the second series, medium strength concrete mixes having nearly the same cube strength at 28 days with natural coarse and fine aggregates (Mix GS1) or recycled coarse and fine aggregates and natural sand (Mix RR1). Natural sand and crushed concrete fines were blended in equal weight proportion to produce the recycled aggregate concrete mix. The water-cement ratios for the natural aggregate concrete and the recycled aggregate concrete were 0.55 and 0.48, respectively. The reduction in the water-cement ratio for the recycled aggregate concrete was required to achieve a similar 28-day cube strength for both types of concrete. A third mix (Mix RRM) in the second series was produced by adding 10% condensed silica fume by cement weight to the Mix RR1.

Standard tests for the hardened concrete were conducted at various ages in accordance with the BS procedures. Tests specimens were cast in steel moulds and demoulded after one day. Moist curing in room temperature of about 28°C was given to these specimens until testing. The drying shrinkage specimens were 100x100x500 mm in size and moist cured for either 3 or 28 days, prior to drying in the laboratory environment of about 28°C and about 80% relative humidity.

Table 2: Summary of test results

Property	Mix / Age(d)	GS	RR	RRF	RRS	GS1	RR1	RRM
Compressive	3	20.3	14.2	18.2	21.4	24.8	25.0	26.0
strength (MPa)	28	33.8	28.0	32.5	35.1	35.5	38.7	39.8
	91	35.3	31.0	38.2	40.3	38.4	43.6	45.6
Flexural	3	-	-	-	-	3.23	3.81	4.47
strength (MPa)	28	4.88	4.15	6.21	6.29	4.26	5.19	5.56
	91	6.50	5.13	8.05	8.16	-	-	-
Tensile	3	-	-	-	-	1.43	1.80	2.25
strength (MPa)	28	2.41	1.93	2.55	2.48	2.55	3.09	3.26
Static	7	26.1	19.7	18.9	19.6	28.5	18.9	22.4
modulus (GPa)	28	29.0	21.5	21.3	22.0	30.4	26.5	26.5
	91	29.1	22.3	22.5	23.0	-	-	-
Pulse velocity	28	4.49	3.96	4.09	4.07	4.44	4.14	4.17
(km/s)	91	4.53	4.04	4.21	4.18	4.47	4.16	4.23
Shrinkage	* 91	204	430	360	306	360	532	508
(microstrain)	@ 91	223	406	-	-	326	408	355

Note: Moist curing period prior to drying - * 3 days; @ 28 days

5. Effects of pozzolanic addition on recycled aggregate concrete

Table 2 summarizes the hardened concrete properties for the mixes upto the age of 91 days. The properties of recycled aggregate concretes in relation to the natural aggregate concrete are given in Table 3. The development of drying shrinkage with time for the concrete mixes after initial moist curing period of 3-days is shown in Fig. 2.

Table 3: Relative properties of recycled aggregate concretes (%)

Property	Mix / Age(d)	GS	RR	RRF	RRS	GS1	RR1	RRM
Compressive strength	3	100	70	90	105	100	101	105
	28	100	83	96	103	100	109	112
	91	100	88	108	114	100	114	119
Flexural strength	3	-	-	-	-	100	118	138
	28	100	85	127	129	100	122	131
	91	100	79	124	126	-	-	-
Tensile strength	3	-	-	-	-	100	126	157
	28	100	80	106	103	100	121	128
Static modulus	7	100	75	72	75	100	66	79
	28	100	74	73	76	100	87	87
	91	100	77	77	79	-	-	-
Pulse velocity	28	100	88	91	91	100	93	94
	91	100	89	93	92	100	93	95
Shrinkage	* 91	100	211	176	151	100	148	141
	@ 91	100	182	-	-	100	125	109

Note: Moist curing period prior to drying - * 3 days; @ 28 days

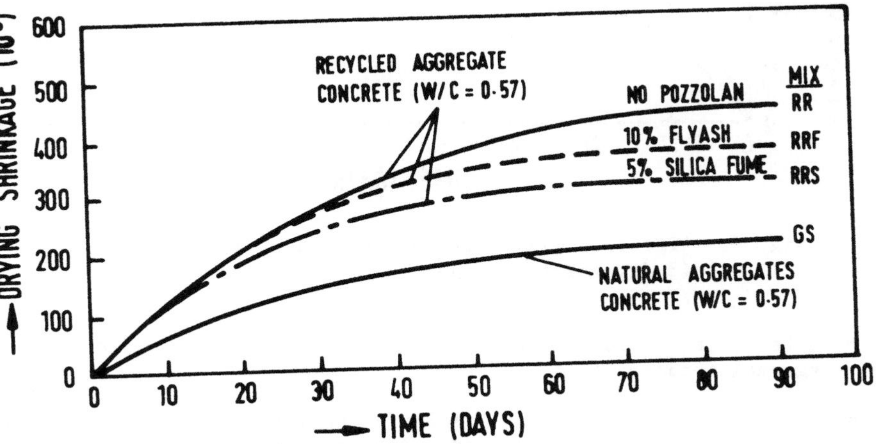

Fig. 2. Effect of pozzolan addition on shrinkage of concrete (3-day moist curing)

From the results the following observations can be made:

(a) addition of either fly ash or silica fume helped to recover the strength of recycled aggregate concrete in compression, tension, and flexure;

(b) 91-day compressive strength of recycled aggregate concrete with 5% silica fume is 14% more than that for the natural aggregate concrete;

(c) 10% addition of fly ash produced only 8% increase in compressive strength at 91 days;

(d) addition of fly ash or silica fume also resulted an improvement in flexural strength of over 25% for recycled aggregate concrete at 28 days and beyond, and reduced in the drying shrinkage;

(e) although the pozzolans improved the strengths and reduced the drying shrinkage of recycled aggregate concrete, the elastic modulus is only marginally improved. Since the elastic modulus of concrete is mainly affected by the aggregate modulus the observations are not surprising. The modulus of elasticity of the recycled aggregate concrete is about 75 to 80% of that of natural aggregate concrete of similar mix proportions;

(f) drying shrinkage of both natural aggregate concrete and recycled aggregate concrete was reduced when the initial moist-curing period was increased from 3 to 28 days.

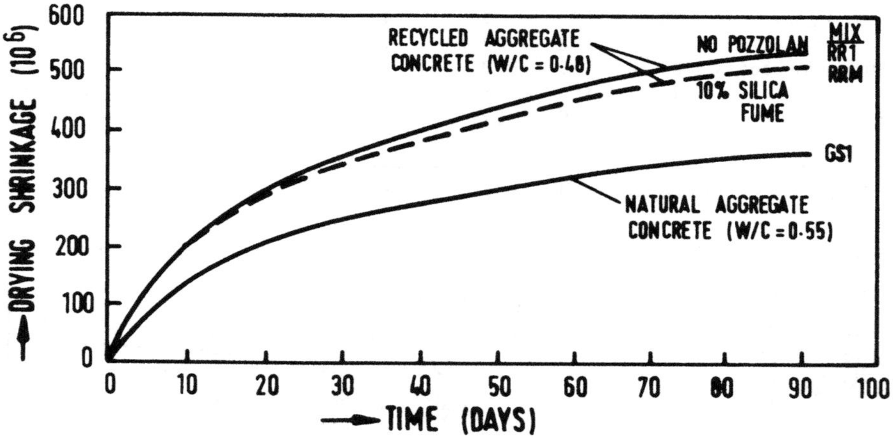

Fig. 3. Effect of mix modification on shrinkage of concrete (3-day moist curing)

6. Effects of mix modification on the properties of recycled aggregate concrete

Hansen and Narud (1983a) and Sri Ravindrarajah and Tam (1985) found that when concrete is crushed in a jaw crusher about 25% by weight of the crusher products is below 5 mm size. Although it is possible to use the concrete fines as fine aggregate, it is uneconomical to generate sufficient volume of concrete fines in order to replace the natural sand fully. Therefore, blending of natural sand with the crushed concrete fines seems to be an economical solution. Furthermore, this method of utilization of concrete fines will reduce the detrimental effects of using recycled aggregates on concrete quality to a certain degree. The addition 10% silica fume by cement weight to the recycled aggregate concrete was also considered.

Table 2 summarizes the results of the properties of the concrete mixes investigated. The properties of recycled aggregate concrete with and without the silica fume in relation to those for natural aggregate concrete are given in Table 3. The development of drying shrinkage with time for the mixes is shown in Fig. 3.

From the results the following observations can be made:

(a) strengths of the recycled aggregate concrete in compression, flexure, and tension are easily recovered by the mix modifications such as a reduction in the water-cement ratio or the addition of condensed silica fume;

(b) for recycled aggregate concrete the modulus of elasticity and pulse velocity are lower than those for the natural aggregate concrete even after the mix modifications were made for the former;

(c) reduction in the water-cement ratio in combination with blending of natural sand with crushed concrete fines in equal weight proportion improved the modulus of elasticity and reduced the shrinkage for the recycled aggregate concrete;

(d) addition of silica fume to the recycled aggregate concrete with reduced water-cement ratio did not produce any significant improvement in either modulus of elasticity or shrinkage.

7. Concluding remarks

Waste concrete from construction industry is either contaminated or free from contamination. If the waste concrete is to be used to produce aggregates for concrete then concrete with a lower degree of contamination is preferred. Studies reported here and elsewhere have shown that the aggregates from waste concrete should be considered as of marginal quality with negative effects on the concrete properties. However, with proper mix modifications and the use of pozzolans such as fly ash or condensed silica fume it may be possible to produce recycled aggregate concrete with acceptable quality for construction similar to lightweight aggregates.

From this investigation it appears that improvement in strength for recycled aggregate concrete can be easily obtained by either a reduction in the water-cement ratio or by the addition of pozzolan. The condensed silica fume showed a better performance than the fly ash. The results of this investigation also show that improvements, in the deformational properties of concrete such as modulus of elasticity and drying shrinkage are difficult to achieve to the same degree as that for strength. However, the quality of recycled aggregate concrete is found to improve considerably when the following measures are taken in combination: (a) use of a lower water-cement ratio: (b) blending of natural sand and crushed concrete fines in equal weight proportion; and (c) addition of fly ash or condensed silica fume. It is known that the pozzolans are capable of modifying the pore structure of the cement paste matrix resulting a decrease in the permeability of concrete. This, in turn, improves the durability of the recycled aggregate concrete. Moist-curing for longer period is essential to improve the volume stability as well as durability of concrete whether it is made with natural or recycled aggregates.

Use of water reducing admixture such as superplasticizer should be seriously considered to improvement the workability of the recycled aggregate concrete without adding more water. The reduction in the water content is helpful in reducing the drying shrinkage of the recycled aggregate concrete.

Acknowledgment

This study was supported by the Ministry of Trade and Industry, Singapore, RDAS Grant C/81/01 on Low Cost Construction Materials.

References

BCSJ (1978) Study on recycled aggregate and recycled aggregate concrete. Building Contractors Society of Japan, Concrete J., Japan, 16(7), 18-31.

Hansen, T.C. and Narud, H. (1983a) Strength of recycled concrete made from crushed concrete coarse aggregate. Concrete International: Design and Construction, 5(1), 79-83.

Hansen, T.C. and Narud, H. (1983b) Recycled concrete and fly ash make calcium silicate bricks. Cement and Concrete Research, 13(4), 507-510.

Hansen, T.C. and Narud, H. (1983c) Recycled concrete and silica fume make calcium silicate bricks. Cement and Concrete Research, 13(5), 626-630.

Hansen, T.C. (1986) Recycled aggregates and recycled aggregate concrete second state-of-the-art report developments 1945-1985. Materials and Structures, 19(111), 201-246.

Ong, K.C.G., Liaw, C.Y. and Tam, C.T. (1985) Fracture energy in recycled aggregate concrete. Proc. of the 11th Conf. on Our World in Concrete and Structures, Singapore, 341-360.

Ong, K.C.G., and Sri Ravindrarajah, R. (1987) Fracture energy of concrete with natural and recycled concrete aggregates. Proc. of Int. Conf. on Fracture of Concrete and Rocks, Houston, USA.

Sri Ravindrarajah, R. and Tam, C.T. (1985) Properties of concrete made with crushed concrete as coarse aggregate. Magazine of Concrete Research, 37(130), 29-38.

Sri Ravindrarajah, R. (1986) Recycling waste concrete for the production of new concrete. Proc. of the 5th Int. Recycling Congress, Berlin, Germany.

Sri Ravindrarajah, R. and Tam, C.T. (1986) Concrete with fly ash or crushed concrete fines or both. Paper presented at the Second Int. Conf. on the Use of Fly Ash, Silica Fume, Slag and Natural Pozzolans in Concrete, Madrid, Spain.

Sri Ravindrarajah, R., and Tam, C.T. (1987a) Recycling concrete as fine aggregate in concrete. The Int. J. of Cement Composites and Lightweight Concrete, 9(4), 235-241.
Sri Ravindrarajah, R., and Tam, C.T. (1987b) Recycled concrete as fine and coarse aggregates in concrete. Magazine of Concrete Research, 30(141), 214-220.
Sri Ravindrarajah, R., Loo, Y.H., and Tam, C.T. (1988) Strength evaluation of recycled aggregate concrete by in-situ tests. Materials and Structures (to appear).

STRENGTHS OF CONCRETE CONTAINING RECYCLED CONCRETE AGGREGATE

T. IKEDA, S. YAMANE and A. SAKAMOTO
Technical Research Laboratory, Takenaka Corporation

Abstract
Concrete wastes have been recycled only for macadams, but they have not yet been utilized for aggregates of concrete.

Therfore the authors conducted studies on the strengths of concrete containing recycled concrete aggregates in order to utilize these wastes.

The concrete with recycled concrete aggregates showed lower compressive strength than ordinary aggregate concrete by 15 percent. The tensile strength, bending strength and shear strength of the concrete were lower than ordinary aggregate concrete by 10, 7 and 32 percent respectively. Similarly, the concrete showed lower elastic modulus and unit weight.

Meanwhile, the development ratios of each strengths with age of the concrete containing recycled concrete aggregates were similar to those of ordinary aggregate concrete. The relationship between the compressive strength and other strengths such as tensile strength were the same as that of ordinary concrete. The relation between the compressive strength and the elastic modulus was also similar to that of ordinary concrete.

Based on these results it was concluded that the concrete containing recycled concrete aggregates could be utilized for general structural concrete.
Key Words: Recycled concrete aggregate, Compressive strength, Tensile strength, Bending strength, Shear strength, Elastic modulus.

1. Introduction

Since the begining of 1970's the demolitions of reinforced concrete buildings have been increasing in the urban areas in Japan, and consequently a large volume of scrapped concrete has been produced. According to a recent survey the total volume of scrapped concrete is estimated at approximately five millions cubic meters, which is almost 10 % of the total volume of new concrete used to construct new reinforced concrete buildings. In most cases these scrapped concretes are dumped for reclamation works. The dumping of scrapped concrete, however, is becoming more difficult because of the shortage of lands for wastes. In addition, problems associated with the dumping can cause serious environmental pollutions.

Therefore, the treatment of scrapped concrete is an important subject in our country.

A part of concrete wastes has been recycled for macadams, but the volume has been extremely small, compared with the total volume of scrapped concrete, and thus a new technique to treat or utilize the scrapped concrete needs to be developed.

A possible mean to utilize the scrapped concrete is to use them as aggregates for new concretes. This mean may solve both the problems of treatment of scrapped materials and reduction of the cost of aggregate in concrete. From this point of view the Building Contractors Society of Japan commenced studies on the use of recycled concrete aggregate in 1974.

This report presents the results of laboratory studies especially on the mechanical properties of concrete containing recycled concrete aggregate.

2. Laboratory study

2.1 Scope of experiment

Compressive, splitting tensile, bending and shear strength, elastic modulus and unit weight were measured on concretes containing recycled concrete aggregates and ordinary aggregates. Table 1 shows the combination of materials used in the test.

Table 1. Combinations of materials used in the test.

Factors	Fine agg.	Coarse agg.	W/C(%)	Slump(mm)
Level 1	NS	NG	45	210
Level 2	NS+CS	CG	60	150
Level 3	—	—	70	—

NS,NG:natural aggregate. CS,CG:recycled concrete aggregate.

2.2 Materials and mix proportions

Recycled concrete aggregates were made by crushing cylinders cast from three kinds of concretes which had different water-cement ratios. Properties of aggregates are shown in Table 2. CS45 or CG 45 in Table 2, for example, denotes recycled sand or gravel made from concrete with W/C of 0.45 respectively. NS or NG indicates river sand or gravel respectively.

The cement used was ordinary portland cement (specific gravity 3.16). Air entraining agent was used only in concretes containing recycled concrete aggregates in order to reduce the water content in these concrete.

2.3 Test methods

Slump, air content and unit weight of fresh concrete were measured. All the specimens for the strength tests were cured in water at 20℃ for 28 days, and after that they were cured in the air at 20℃ and 80 % relative humidity until testing.

Table 2. Properties of aggregates.

Kind of aggregate		Specific gravity	Absorption (% by weight)	Fineness modulus
Fine aggregate	NS	2.59	1.35	2.70
	CS45	2.02	11.63	3.67
	CS55	2.05	11.12	3.82
	CS68	2.03	11.75	3.83
Coarse aggregate	NG	2.63	0.85	7.35
	CG45	2.31	5.72	7.55
	CG55	2.32	5.69	7.63
	CG68	2.33	5.72	7.14

Table 3. Mix proportions and Test results of fresh concrete.

No	Kind of aggre-gate	W/C (%)	Nominal Slump (mm)	Nominal Air (%)	Weight (kg／m³) Cement	Water	Fine agg.	Coarse agg.	Slump (mm)	Air (%)	Unit Weight (kg／m³)
1	NS・NG	45	210	1	476	214	725	907	208	1.2	2378
2	NS・CG	45	210	4	476	214	749	707	215	4.4	2205
3	NS+CS45 ・CG45	45	210	4	476	214	666	707	211	5.2	2128
4	NS・NG	60	210	1	340	204	852	918	210	2.2	2327
5	NS・CG	60	210	4	340	204	888	708	202	6.2	2146
6	NS+CS55 ・CG55	60	210	4	340	204	796	708	215	6.2	2102
7	NS・NG	70	210	1	290	203	909	905	214	1.7	2330
8	NS・CG	70	210	4	290	203	956	690	215	6.3	2145
9	NS+CS68 ・CG68	70	210	4	290	203	853	690	217	5.8	2071
10	NS・NG	45	150	1	416	187	694	1060	158	1.1	2411
11	NS・CG	45	150	4	416	187	730	829	171	5.6	2193
12	NS+CS45 ・CG45	45	150	4	416	187	650	829	156	5.8	2135
13	NS・NG	60	150	1	297	178	782	1094	151	2.0	2374
14	NS・CG	60	150	4	297	178	855	831	167	6.6	2136
15	NS+CS55 ・CG55	60	150	4	297	178	565	831	135	6.2	2120
16	NS・NG	70	150	1	254	178	816	1097	150	1.7	2370
17	NS・CG	70	150	4	254	178	909	818	170	6.4	2153
18	NS+CS68 ・CG68	70	150	4	254	178	811	818	140	6.0	2089

3. Test results and discussion

3.1 Fresh concrete

Test results of fresh concretes are shown in Table 3. Variations in slump values of concretes containing recycled concrete aggregates were larger than those of ordinary concrete. This seems to be due to the change of the size of recycled concrete aggregate during mixing, that is, crushing of recycled concrete aggregate.

The larger the content of recycled concrete aggregate was, the smaller the unit weight of the concrete was. This tendency obviously attributes the difference in the specific gravities of recycled concrete aggregate and ordinary aggregate.

3.2 Compressive strength

Fig. 1 shows the relationships between compressive strength and age. Compressive strengths of the concretes with recycled concrete aggregates were 15 to 40 % lower than those of ordinary concrete at each age, but the development of compressive strength with age was similar to that of ordinary concrete.

Fig. 2 indicates the relationship between compressive strength and water-cement ratio. In all cases of combination of aggregates, slump and age, the increase in the compressive strength with the decrease in water-cement ratio was observed. The strength-reducing effect by use of recycled concrete aggregates seems greater in the lower water-cement ratios.

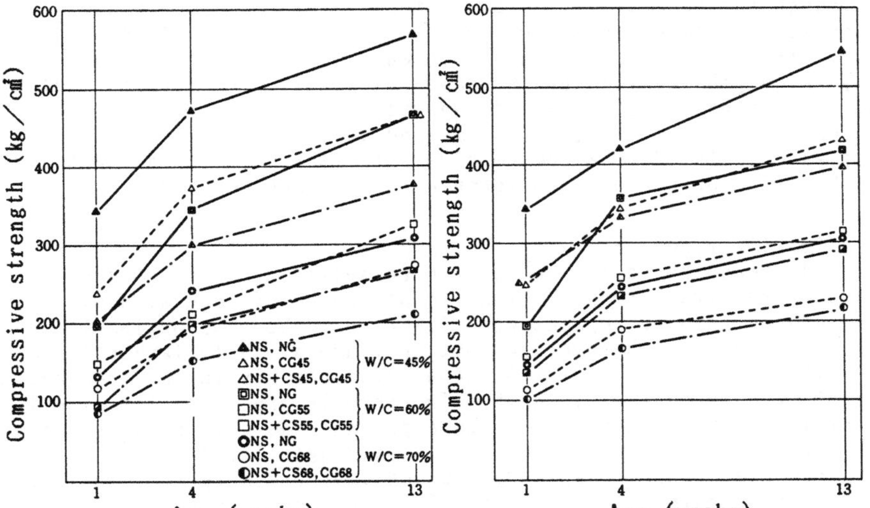

(a) Slump 210 mm (b) Slump 150 mm

Fig. 1. Relationships between compressive strength and age.

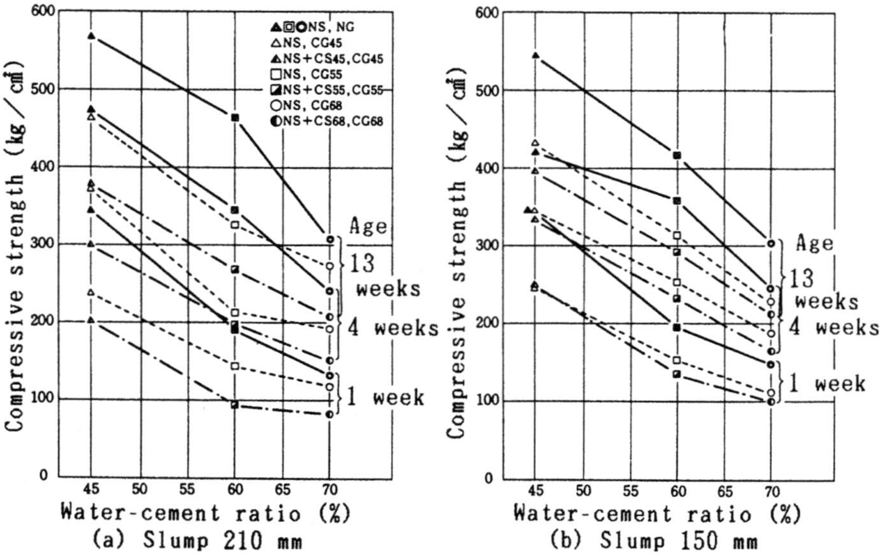

Fig. 2. Relationships between compressive strength and water-cement ratio.

3.3 Elastic modulus

Fig. 3 gives the test results of elastic modulus at compression. Although the reduction in elastic modulus of concretes containing recycled concrete aggregates was more marked than that in compressive strength, the relationship between elastic modulus and compressive strength seems to be the same. Actually, the measured values of elastic moduli of recycled aggegate concretes were in the range between 1.9 and 2.3×10^5 kg／cm², and this range coincides with values calculated from compressive strengths based upon the formula in Fig. 3. Origin of this formula is recommended by A.C.I.

3.4 Tensile strength

Fig. 4 shows the relationship between splitting tensile strength and compressive strength. Tensile strengths of concretes with recycled concrete aggregates were lower than those of ordinary concrete by 25 % on the average, but they are slightly larger than past results obtained by Sen or Akazawa reference missing at the same level of compressive strength. Therfore in practice, the experimental formula by Akazawa may be applicable to obtain the relationship between tensile and compressive strength for concrete containing recycled concrete aggregate as well as ordinary oncrete. Also, it may be possible to use ft=fc/12 in the compressive strength range of 200 to 400 kg／cm² for every kind of aggregates.

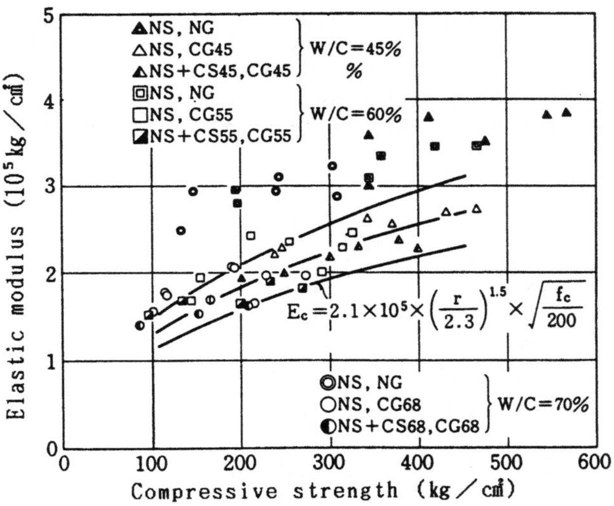

Fig. 3. Relationship between elastic modulus and compressive strength.

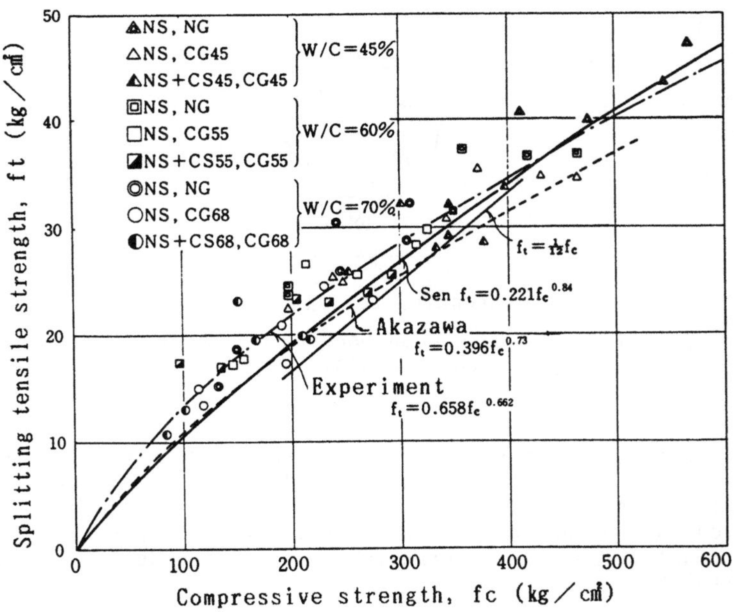

Fig. 4. Relationship between splitting tensile strength and compressive strength.

3.5 Bending strength and shear strength

As shown in Fig. 5, the correlation between bending strength and compressive strength is not so clear as in the case of tensile strength. Reduction of bending strength of concrete with recycled concrete aggregates was approximately 20 %, which is the smallest in the four strengths.

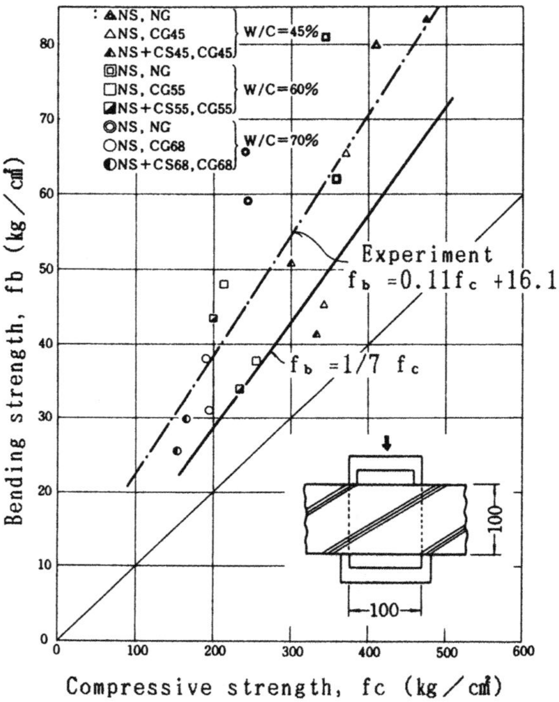

Fig. 5. Relationship between bending strength and compressive strength.

The relationship between shear strength and compressive strength, as shown in Fig. 6, is also not so clear, but the reduction of shear strength by using recycled concrete aggregates is approximately 40 %, which is the largest. In both cases of bending strength and shear strength, they can be practically estimated as 1/7 of compressive strength.

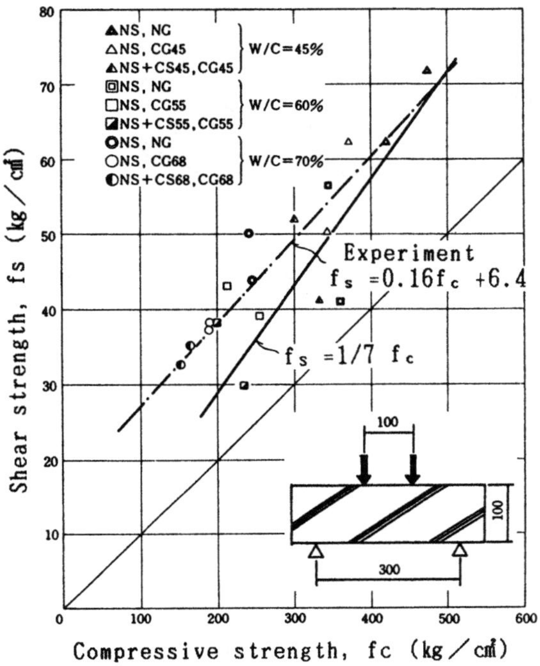

Fig. 6. Relationship between shear strength and compressive strength.

3.6 Unit weight

Unit weight of concrete was determined with specimens for tensile strength at age of 13 weeks. Specimens were cylinders of 100 mm diameter and 200 mm long. Fig. 7 shows relationship between unit weight and compressive strength. It can be seen that the compressive strengths decrease with the decrease in the unit weight, regardless of water-cement ratio and aggregates. The reduction ratio of unit weight by using recycled concrete aggregate is about 10 %.

3.7 Applicability of recycled concrete aggregate

The average ratios of compressive, tensile, and shear strength of concrete with recycled concrete aggregates to those of ordinary concrete are listed in Table 4. Corrected data in this table were determined based upon the difference in the air content. In other words, 1 % difference in the air content produces 4 % differences in the strength. Since air-entraining concrete is commonly used in Japan, the comparisons should be made on the basis of equal level of air content.

First of all, on original data the reduction in compressive and tensile strengths of concrete with recycled concrete as coarse ag-

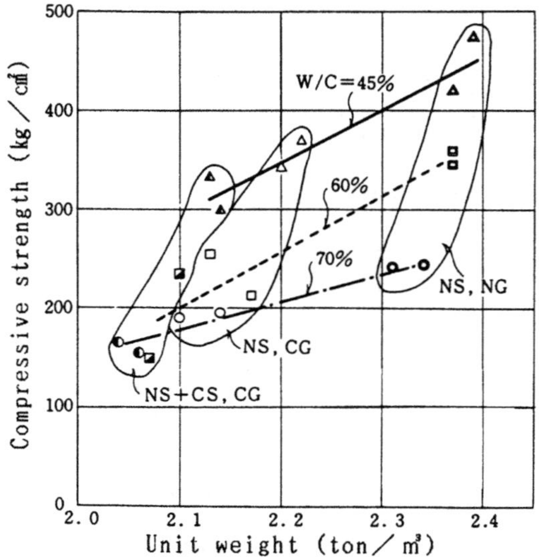

Fig. 7. Relationship between unit weight and Compressive strength.

Table 4. Strength ratios of concrete with recycled concrete aggregate to those of ordinary concrete.

	Kind of aggregate	Strength ratio, original data		Strength ratio, corrected data	
Compressive strength	NS・CG	77	72	92	86
	NS+CS・CG	66		79	
Splitting tensile strength	NS・CG	79	76	94	90
	NS+CS・CG	72		86	
Bending strength	NS・CG	84	78	101	93
	NS+CS・CG	71		85	
Shear strength	NS・CG	61	57	74	68
	NS+CS・CG	52		62	

gregate were approximately 20 %, and reduction in bending and shear strength were 15 and 40 % respectively. In addition, further reduction of approximately 10 % would be produced for all strengths if recycled concrete aggregate is also used as fine aggregate. On the other hand, on corrected data the reduction ratios are smaller than on the original data by approximately 15 %. Thus the strength reduction cause by the use of recycled concrete aggregate is small except for shear strength; 0～8 % for NS・CG, and 15～20 % for NS+CS・CG. Then it may be considered that concrete with recycled concrete aggregate can be applied to structural concrete with proper consideration on the shear strength, for example, rainforcement for shear force.

4. Conclusions

The matters that can be concluded based on the above experiments are as follows:

(1) Reductions in strengths caused by the use of coarse recycled concrete aggregate are estimated at 8 % for compressive strength, 6 % for tensile strength, 0 % for bending strength and 26 % for shear strength, if the effect of difference in the air content was considered. If recycled concrete aggregate is used also as fine aggregate, further 10, 15 % reductions may be produced.

(2) The characteristics of compressive strength, tensile strength and elastic modulus of concrete containing recycled concrete aggregate with change of age or water-cement ratio are the same as for those of ordinary concrete.

(3) The relationship between compressive strength and other types of strength such as tensile strength are similar in both concretes with recycled concrete aggregates and natural aggregates.

(4) Although elastic moduli of concretes containing recycled concrete aggregate are lower by 30 to 50 % than those of ordinary concrete, they show good fitting with formula recommended by A.C.I.

(5) Unit weights of concretes with recycled concrete aggregate are approximately 10 % smaller than those of ordinary concrete. However, a good relation was observed between unit weight and compressive strength of concretes regardless of differences in water-cement ratio or aggregate.

In conclusion, the recycled concrete aggregate could be utilized for structural concrete, if suitable considerations are made on the content of recycled concrete aggregate and on the design for shear force applied to the concrete.

References

Sen, B.R. and Bharara, A.L. (1961) A new indirect tensile test for concrete. Indian Concrete J., 35, 3, Mar.

Akazawa, T. (1943) A new test to estimate inner stress of concrete by compression (on a splitting tensile test). J. of the Japan Society of Civil Engineers, 29, 11, Nov.

PROPERTIES OF CONCRETE PRODUCTS CONTAINING RECYCLED AGGREGATE

MASAFUMI KIKUCHI Department of Architecture, University of Meiji
TAKESHI MUKAI Department of Architecture, University of Meiji
HARUO KOIZUMI KOIZUMI Building Materials Corporation

Abstract

For the purpose of practical application of recycled aggregates, concrete products containing recycled aggregates were manufactured for trial. These were manufactured by practical equipments in a concrete product factory, and the qualities of these products were compared with those of products made of natural aggregate concrete.

As a result, it was confirmed that it does not matter in practical application in case of replacement ratio by recycled aggregates within 30 per cent.

Key words: Original concrete, Recycled aggregate, Recycled aggregate concrete products, Pressed cement roof tiles, Hollow concrete brocks, Reinforced concrete gutters, Reinforced concrete curb gutters, concrete curbs.

1. Introduction

We have been investigating the application of waste concretes discharged from demolishing of reinforced concrete structures as aggregates for new concretes since 1974.

As a result of obtained by the authors, it is observed that the deterioration in qualities of recycled aggregate concretes were proportional to the replacement ratio by recycled aggregates.

In case of concretes containing recycled aggregates under 30 percent by volume of the whole aggregate, a large difference between natural aggregate concretes and recycled aggregate concretes was not recognized. Moreover, effects of steam curing on recycled aggregate concretes were similar to those for natural aggregate concretes.

Therefore, it can be said that the concrete with partially replaced natural aggregate by recycled aggregate is able to apply sufficiently for structural concrete. However, in Japan, some period is necessary until recycled aggregate concretes are put into pratical application for structural concrete, and it is necessary to accumulate the results by many pilot studies such as simple structures or concrete products. From this view point, small size concrete products for public construction were manufactured using practical equipment in a concrete products factory, and the fundamental properties of them and feasibility of an application of this type of concrete products were invesigated.

2. The outline of recycled aggregate in this investigation

2.1 The original concretes and manufacturing method of recycled aggregates

Two types of concrete containing natural aggregate were prepared to produce recycled aggregates. The outline of manufacturing method of recycled aggregates are shown in Table 1.

2.2 Type of aggregates and its properties

Type of aggregates used to manufacture concrete products and their properties are show in Table 2.

Table 1 The outlines of recycled aggregates prepared in this investigation

Series	Original Concrete	Production of recycled aggregate
I	Natural sand and gravel concrete, steam cured concrete manufactured by precast concrete factory. Compressive strength at crushing of original concrete:220 ～375kgf/cm²	Original concrete was crushed by jaw crusher and screened to required aggregate gradation at the age of 14 months.
II	Natural sand and gravel concrete, normal cured concrete manufactured by ready mixed concrete factory. Compressive strength at crushing of original concrete:240 ～350kgf/cm²	Original concrete was crushed by jaw crusher, screened and classified into coarse and fine aggregate at the age of 35 days.

Table 2 Main properties of aggregate prepared in this investigation

Series	type of aggregate			Maximum size (mm)	FM	Specific gravity	Absorpion (%)	Unit weight (kg/l)	Ratio of absolute volume(%)
I	fine	recycled	CS_1	(5)	3.66	2.15	8.9	1.41	65.6
			CS_2	(10)	4.28	2.20	7.7	1.36	62.0
		natural	NS_1	(5)	2.82	2.51	1.8	1.61	64.3
			NS_2	(10)	3.21	2.54	1.6	1.65	65.1
II	fine	recycled	CS	(5)	3.48	2.05	10.5	1.48	72.2
		natural	NS	(5)	2.79	2.54	0.8	1.64	64.7
		ash[1)]	CA	—	1.66	1.76	3.4	0.89	47.9
	coarse	recycled	CG	20	6.51	2.22	7.0	1.31	59.0
		natural	NG	20	6.69	2.63	0.6	1.51	57.3

1) Clinker ash discharged from electric power station

3. Concrete products manufactured for trial in this investigation

3.1 Type of concrete products, shape and size.

Pressed cement roof tiles, Hollow concrete blocks(above,series I), Reinforced concrete gutters, Reinforced concrete curb-gutters and Concrete curbs(above,series II) were manufactured for trial.The outline of these concrete products are shown in Table 3 and Fig.1a)～e).

Table 3 The outlines of concrete products manufactured for trial

Series	Type of concrete products	JIS[1] NO.	Remarks
I	Pressed Cement Roof Tiles : S type Hollow Concrete Blocks : C type	JIS A 5402 JIS A 5403	See Fig.1a) See Fig.1b)
II	Reinforced Concrete Gutters : 240 type Reinforced Concrete Curb-Gutters:250B type Concrete Curbs for site frontage A type for roadway A type	JIS A 5305 JIS A 5306 JIS A 5307	See Fig.1c) See Fig.1d) See Fig.1e) See Fig.1e)

1) Japanese Industrial Standards

3.2 Type of concrete mixes prepared to manufacture concrete products

Varying the replacement ratio by recycled aggregate, 6 types and 4 types of concrete mixes are prepared in series I and II respectively. These mix proportions are shown in Table 4.

Table 4 The outlines of concrete mixes prepared to manufacture concrete products

Series and Type of mixes		Type of aggregate used (by volume : %) coarse NG	coarse CG	fine NS	fine CS	fine CA	Mix proportion W/C (%)	Cement content (kg/m³)	Water content (kg/m³)	Concrete products manufactured for trial
I	NS_1	–	–	100	0	–	45	665	299	·Pressed Cement Roof Tiles
	NS_1CS_1	–	–	50	50	–		674	303	
	CS_1	–	–	0	100	–		646	291	
	NS_2	–	–	100	0	–	32	250	100	·Hollow Concrete Blocks
	NS_2CS_2	–	–	50	50	–		250	100	
	CS_2	–	–	0	100	–		257	82	
II	NG · NS	100	0	100	0	0	50	320	160	·Reinforeced Conrete Gutters
	$N_{70}C_{30}$	70	30	70	30	0		340	170	·Reinforeced Conrete Curb-Gutter
	N.C.CA	70	30	70	15	15		350	175	·Concrete Curbs
	CG · NS	0	100	100	0	0		350	165	

3.3 Method of mixing, forming and curing

Method of mixing, forming and curing were the same way as the method adopted usually in that concrete products factory.

4. Examination items and testing methods

Strength, absorption, permeability, drying shrinkage and carbonation were measured. The outline of testing methods of various concrete products are shown in Table 5 and Fig. 2.

5. Experimental results

5.1 Experimental results of Series I

Pressed cement roof tiles and hollow concrete blocks were manufactured from 3 types of concrete mixes respectively.
Main test results are shown in Table 6 ～ 7 and Fig. 3.

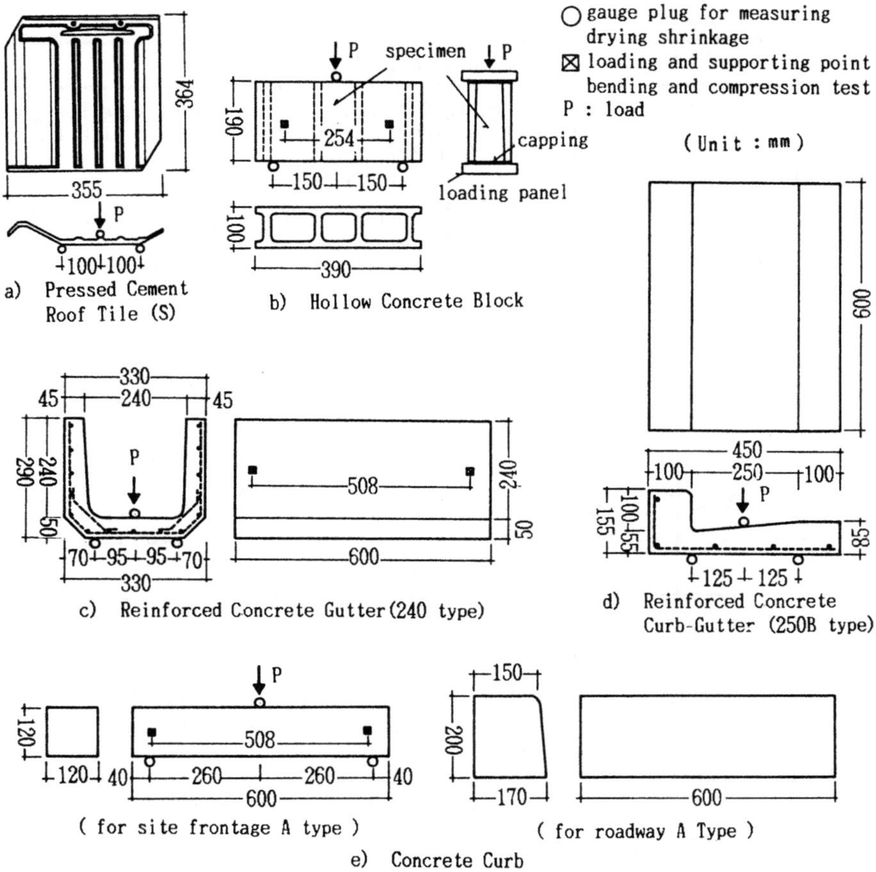

Fig.1 The outlines of shape, size and testing method on concrete products manufactured for trial.

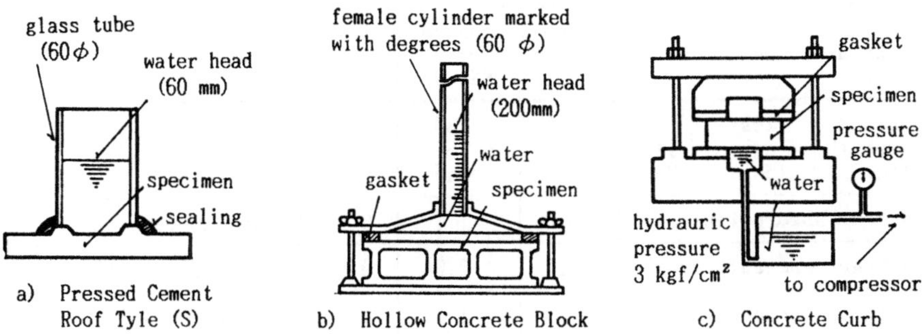

Fig.2 The outlines of testing method on permeability

Table 5 The outlines of testing method

Test items	The outlines of testing method
Bending strength	· The method provided with the JIS concerned.(see Fig.1)
Compressive strength	· Hollow Concrete Blocks : The method provided with the JIS concerned.(see Fig.1 b) · Concrete Curbs : Specimen, the broken pieces of bending specimens.
Absorption	· Pressed Cement Roof Tiles and Hollow Concrete Blocks : The method provided with the JIS concerned. · Concrete Curbs : Specimens that cut down Concrete Curbs for site frontage to shape of 12×12×5cm. These were immersed in water for 7 days.
Permeability	· Pressed Cement Roof Tiles : Specimens that cut down Pressed Cement Roof Tiles to shape of 14×14cm. These were tested by the method shown in Fig.2 a) · Hollow Concrete Blocks : The method provided with the JIS concerned. (see Fig.2 b) · Concrete Curbs : Specimens, the same as absorption. These were tested by the method provided with JIS A 1404 shown in Fig.2 c). Hydraulic pressure:3 kgf/cm^2-1hour
Drying shrinkage	· Length change of specimens were measured by Huggenberger strain gauge.(see Fig. 1)
Carbonation	· Curing : exposed outdoor. · Judgement : spraying by phenolphthalein.

(1) Pressed cement roof tiles(See Table 6)

(a) Mix proportion of mortars

Mortars that have a similar consistency were obtained by the following mix proportion : that is water-cement ratio: o.45, cement content about 660 kg/m^3 and cement to fine aggregate ratio about 2.3.

(b) Formability and appearance

A large difference in appearance between NS_1 type (containing natural aggregate) and NS_1CS_1 type (containing natural aggregate and recycled one) could not be recognized. On the other hand, CS_1 type (containing recycled aggregate) showed a somewhat more rough surface than that of the former two.

Table 6 Main test results of Pressed Cement Roof Tiles

Examination items			Type of concrete mixes			Values provided with the JIS concerned and remarks
			NS_1	NS_1CS_1	CS_1	
bending strength (kg)	test age (days)	7	146	135	97	Above 150 kgf
		14	-	156	119	
		28	170	145	111	
24 hour absorption(%)			5.6	7.7	9.6	Under 10 per cent
24 hour permeability(%)			0.3	0.6	0.6	See Fig.2 a) [1)]
depth of carbonation(mm)			1.1	1.1	1.3	At 6 months[2)]

1) Water head Pressure : 6gf/cm^2 2) Exposed outdoor

(c) Bending strength

Bending strength of NS_1CS_1 and CS_1 type were from 110 to 150 kgf. These values were lower compared with the specified JIS values and those of NS_1 type.

From this result, it is necessary to re-examine particle size of aggregate, mix proportion, method of forming, curing and so on.

(d) Absorption and permeability

Values of absorption of NS_1CS_1 and CS_1 type at 24 hours were 7.7 and 9.6 percent respectively, and these values are conformed to the values difined by JIS A 5402 (under 10 percent). A permeation of water to reverse side was not observed in any types.

(e) Carbonation

Carbonation of roof tiles containing recycled aggregate showed a tendency to increase over a long period of time.

(2) Hollow concrete blocks(See Table 7)

(a) Mix proportion

The same mix proportion could be adopted to manufacture NS_2 and NS_2CS_2 type. While, for CS_2 type, water content decreased about 20 percent compared with the former two because of its rough particle size.

(b) Formability and appearance

There were no difference in consistency in mortars of NS_2 and NS_2CS_2 types, and the required time to compact by vibrating were 2.5 seconds. For CS_2 type, it was necessary to increase the required time of compacting to 3.5 seconds. A sligthly difference was recognized in appearance between NS_2 type and NS_2CS_2 type. On the other hand, CS_2 type showed a remarkably rougher surface compared with the former two even after increased vibrating time.

Table 7 Main test results of Hollow Concrete Blocks

Examination items			Type of concrete mixes				Values provided with the JIS concerned and remarks
			NS_2	NS_2CS_2	CS_2 2.5[1)]	CS_2 3.5[2)]	
compressive strength (kgf/cm^2)	test age (days)	7	72.1	78.0	62.8	76.6	Above 60 kgf/cm^2
		14	72.0	86.5	69.4	77.0	
		28	74.8	88.1	—	78.4	
bending strength (kgf/cm^2)		7	26.7	29.7	28.1	27.2	Tested after soaking in water(2 hr)
		28	21.0	30.8	—	26.5	
air dried specific gravity			2.00	1.97	—	1.84	Cured in-doors
24 hour absorption (%)			13.6	16.2	—	21.0	Immersed in water
permeability [3)]			7'13"	2'53"	—	0'07"	(minutes- seconds)
depth of carbonation (mm)			4.5	4.1	—	3.8	At 6 months

1), 2) Required time for vibrating to form Hollow Concrete Blocks.
3) Required time till water level that shown in Fig.2 b) fall down to 10 cm level. water head Pressure : $20 gf/cm^2$

(c) Compressive strength

Compressive strength of various types of blocks were from 63 kgf/cm^2 to 88 kgf/cm^2, and these values conformed to the specified JIS values(above 60 kgf/cm^2). For this reason, it can be said that replacement ratio by recycled aggregate is not a serious problem with regard to strength.

(d) Bulk specific gravity

Bulk specific gravity of blocks containing recycled aggregate were from 1.84 kg/l to 1.97 kg/l, these were slightly lighter than that of conventional type.

(e) Absorption and permeability

The amount of absorption of blocks containing recycled aggregate increased from 13 percent to 21 percent as replacement ratio by recycled aggregate increases.

For permeability, CS_2 type showed an instant water permeation. This type of blocks can not be expected to be watertight without adjusting particle size of aggregate and mix proportion.

(f) Drying shrinkage (See Fig. 3)

Drying shrinkage of CS_2 type showed a value twice as much as these of other types. The value of drying shrinkage had a tendency to increase by repetition of immersing in water and drying in air.

(g) Carbonation

There were no difference in depth of cabonation on various type of blocks. However it was observed that depth of carbonation decreased as replacement ratio by recycled aggregate increased. A similar tendency was observed in series II .

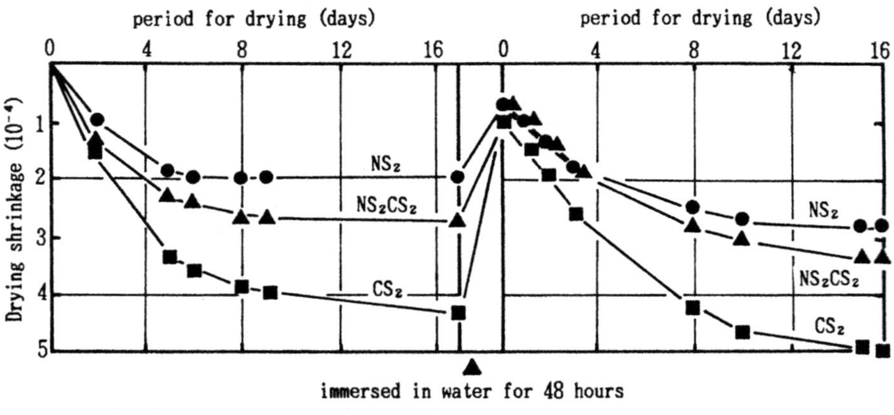

Fig.3 Drying shrinkage of various Hollow Concrete Blocks.

5.2 Experimental results of series II

Reinforced concrete gutters (240 Type), Reinforced concrete curb-gutters(250 B Type) and Concrete curbs(two types) were manufactured from 4 types of concrete mixes that varied the replacement ratio by recycled aggregate.

Main test results on these concrete products were shown in Table 8 and Fig.4 ~ 6.

(1) Mix proportions of concretes

For recycled aggregate concretes which replaced natural aggregate by recycled aggregate in the range of 15 ~50 percent by volume, it was necessary to make water content, cement content and fine aggregate ratio increase. The amount of these increases were 5 ~15 kg/m^3, 10~ 30 kg/m^3 and 1 ~ 2 percent respectively.

(2) Formability and appearance

Under the before mentioned mix proportions, concretes were mixed, packed by dividing in two layers and vibrated by table vibrator for 20 seconds in each layer.

There were no difference caused by type of aggregates or replacement ratio in formability and appearance of concrete products. It is assumed that these results were obtained by adjusting the Finess modulus of aggregate, especially at the fine aggregate.

(3) Bending strength (See Fig. 4)

Bending strength of reinforced concrete gutters (240 Type), reinforced concrete curb gutters(250B type) and concrete curbs(for site frontage Atype) of NG · NS type concrete products were 2175kgf, 2430 kgf and 1163kgf respestively. While, that of recycled aggregate type concrete products decreased in inverse proportion to replacement ratio by recycled aggregate. This tendency was remarkable in plain type products, that is to say, site frontage A Type and CG · NS type that replaced all of coarse aggregate by recycled aggregate.

It can be said that these tendencies were caused by the weakness of recycled aggregates and cracks occured in the surface of concrete.

However, the values of bending strength on recycled aggregate type concrete products were higher by 1.2 to 1.6 times than that of the each defined JIS values.

(4) Compressive strength

Compressive strength of site frontage A Type that were tested by using the broken pieces of bending specimens showed a similar tendency such as bending strength.

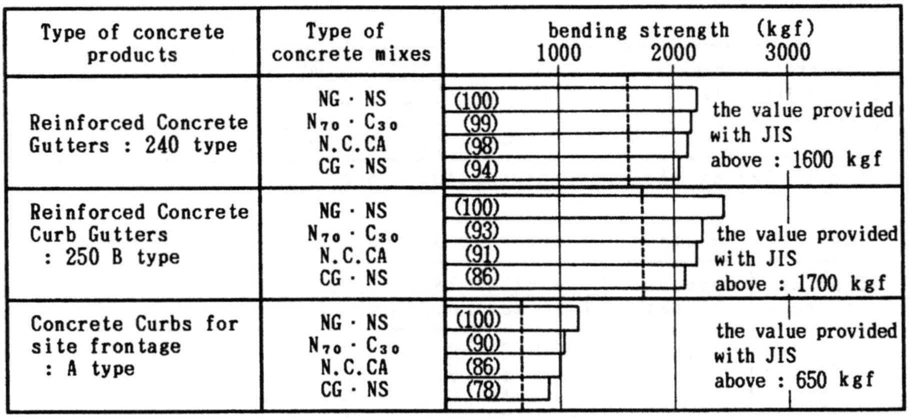

· values in parenthesis show an index number on concrete products containing recycled aggregate to that of NG · NS

Fig.4 Bending strength of various concrete products

(5) Absorption and Permeability(See Fig. 5)

Values of absorption and permeability on concrete products containing recycled aggregate increased in direct proportion to replacement ratio by recycled aggregate. This tendency mainly depends on higher absorption of recycled aggregate.

(6) Drying shrinkage(See Fig. 6 and Table 8)

Values of drying shrinkage of reinforced concrete gutters containing recycled aggregate were $4.9 \sim 5.2 \times 10^{-4}$, and these values were higher by 30 percent to 40 percent compared with that of natural aggregate type. In case of site frontage A type made of plain concrete, drying shrinkage progressed somewhat faster than that of reinforced type. Width and length of craks occurring in surface were largest in CG · NS type that replaced all of coarse aggregate by recycled aggregate. There were little difference in the behavior of cracking for other 3 types.

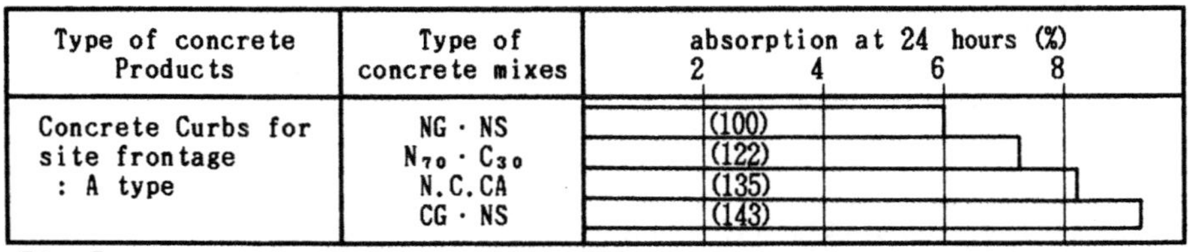

Fig.5 Absorption of Concrete Curbs for site frontage A type

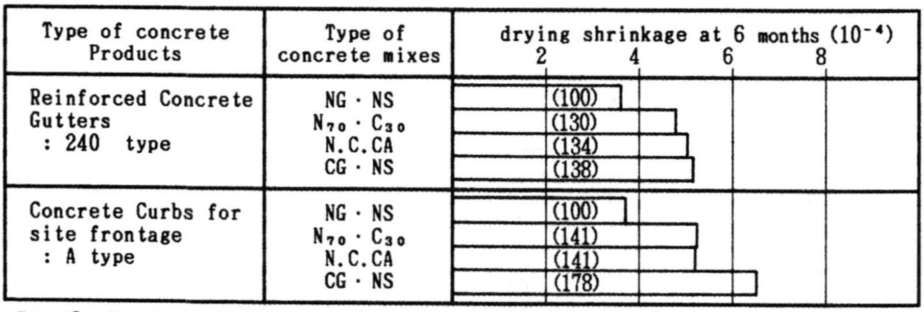

Fig.6 Drying shrinkage of various concrete products

Table 8 The states of cracking of concrete curbs(at 2 months)

Type of concrete	number of cracks	width(10^{-3}mm)			length(mm)			area of cracks($10^{-3}mm^2$)
		min	max	ave	min	max	ave	
NG · NS	12	5	20	8.3	3	22	7.8	390
N70 C30	12	5	10	6.3	5	19	9.8	368
N.C.CA	7	5	10	6.4	5	10	6.4	347
CG · CS	11	5	20	9.5	4	23	13.4	704

· Cracks occured on the bottom at placing of concrete were measured.
· Area of cracks=number of cracks ×average width ×average length/2

(7) Carbonation

Progress of carbonation of recycled aggregate types was generally slow compared with that of natural gravel type. It can be guessed as following in regard to this reason. That is to say, the recycled aggregates prepared in this investigation were supplied as soon as orignal concrete had been crushed, and these original concrete were not entirely carbonated. Therefore the concretes containing recycled aggregates became richer in alkalinity than natural aggregate concretes.

6. CONCLUSIONS

The investigations on application of recycled aggregates to concrete products are summerized as follows.

(1) The particle size of recycled aggregate have a direct influence upon properties of concrete products, for instance, formability, appearance, water-tightness and so on. Therefore, it is necessary to adjust the suitable grade by blending recycled aggregate and natural aggregate or fine particles such as clinker ash.
(2) In case of using suitably graded recycled aggregates, and making a few modification on mix proportions, concrete products can be manufactured by the same procedure as a conventional one.
However, qualities of these concrete products are lowered as a replacement ratio by recycled aggregates increases.
(3) As for the range of replacement ratio which does not matter in the practical application, it is recommend the value within 30 percent.

Reference

Kikuchi Masafumi and Mukai Takeshi (1981) " Study on the Properties of Concretes Containing Recycled Aggregate " CAJ Review, pp.84-87.
Kikuchi Masafumi and Mukai Takeshi (1983) " Study on the Properties of Recycled Aggregate and Recycled Aggregate Concretes " CAJ Review, pp.92-95.
Kikuchi Masafumi and Mukai Takeshi (1984) " The Effect of the Ratio of Replacement by Recycled Aggregate on Properties of Sand-Crushed Stone Concrete " CAJ Review, pp.236-239.
Kikuchi Masafumi, Mukai Takeshi and Koizumi Haruo (1987) " Study on the Properties of Concrete Manufactures Containing Recycled Aggregate " Summary of Technical Paper of Annual Meeting. Architectural Institute of Japan, PP.37-38.

STRENGTH OF RECYCLED CONCRETE MADE FROM COARSE AND FINE RECYCLED CONCRETE AGGREGATES

TORBEN C. HANSEN AND MARIUS MARGA
Building Materials Laboratory, The Technical University of Denmark

Abstract
Compressive strength of hardened concretes made with both coarse and fine recycled aggregates was studied as a function of compressive strength of original concretes from which the recycled aggregates were derived. It was found that based on equal slump, the water requirement of concrete made with both coarse and fine aggregates was 14 percent higher than that of control concretes made with natural sand and gravel. Based on equal water-cement ratios the reduction in compressive strength of recycled concretes was an average of 30 percent. However, the water-cement ratio of such concretes was very difficult to control because common standard testing methods for determining when fine recycled aggregate is in a saturated and surface-dry condition are inappropriate and highly inaccurate. As a consequence of the findings, it cannot be recommended to attempt to produce quality concrete with fine aggregate which is derived from crushed concrete.

1. Introduction

Critical shortage of natural aggregate for production of concrete is developing in many urban areas all over the world. At the same time, increasing quantities of demolished concrete from deteriorated and obsolete structures are generated as waste material in the same areas. It would be desirable if demolished concrete could be processed to yield quality aggregate for production of new concrete. However, there are problems and uncertainties both concerning mechanical properties and durability of recycled concrete. A state-of-the-art report on recycled aggregates and recycled concrete has earlier been prepared by Hansen for RILEM Technical Committee 37-DRC on demolition and recycling of concrete /1/.

Hansen and Narud /2/ have earlier shown that the compressive strength of recycled concrete which is produced with *crushed coarse recycled concrete aggregate and natural sand* is largely controlled by two factors, i.e. the water-cement ratio of the original concrete and the water-cement ratio of the recycled concrete when other factors are essentially identical.

In the present project the authors have studied the compressive strength of recycled concrete which is produced with *both coarse and fine recycled concrete aggregates*. The same cement, original aggregates, and concrete mix proportions were used in this investigation as in the investigation which led to the report published earlier by Hansen and Narud /2/. Although somewhat different curing conditions were used in order to age the concretes, the results of the two investigations can be compared, see Tables 3 and 4.

2. Experimental Design

A high-strength, a medium-strength, and a low-strength concrete were produced from natural sand and gravel by varying the water-cement ratio and keeping other factors essentially identical. After 204 days of curing in water at 20 C, the compressive strength of the three original concretes was determined, and the concretes were passed through a jaw crusher. The crusher products were screened into four size fractions and recombined into three grades of recycled aggregate, all having approximately the same particle grading as the natural materials from which the original concretes were produced.

New high-strength, medium-strength, and low-strength concretes were produced from each of the three grades of recycled aggregate. Compressive strengths of original and recycled concretes were determined and compared after 14 and 204 days of curing in water at 20 C.

3. Preparation and Properties of Materials

3.1 Portland Cement

A high-early strength Portland cement was used, similar to an ASTM Type III cement in composition and properties.

3.2 Natural Aggregates

A partly siliceous, partly calcareous marine gravel and a natural sand of glacial origin were used for production of all original concretes. The sand was graded in BS Zone 2. Density and water absorption of natural coarse and fine aggregate are presented in Table 1 as determined by means of ASTM standard testing methods.

3.3 Recycled Aggregates

Recycled coarse and fine aggregates were prepared by the crushing of original high-strength, medium-strength, and low-strength concretes in a laboratory jaw crusher which was set at an opening of 25 mm with the jaws in a closed position. The crusher products were screened into four size fractions (30-20 mm, 20-10 mm, 10-5 mm, and screenings < 5 mm) and recombined into three grades of coarse and fine aggregate, all having approximately the same grading as the natural material from which the original concretes were produced. Density and water absorption of the recycled aggregates are presented in Table 1.

In the process of carrying out the experiments it became evident that the use of ASTM C128, "Standard Test Method for Specific Gravity and Absorption of Fine Aggregate" to assess when fine recycled aggregate is in a saturated and surface-dry condition is inappropriate and highly inaccurate. Thus the determination of the moisture content of fine recycled aggregates and therefore the effective water-cement ratio of the recycled aggregate concretes are somewhat uncertain. These uncertainties are reflected in the compressive strength test results of the recycled aggregate concretes which were obtained in this investigation. The fact that it is difficult to control the effective water-cement ratio of such recycled concretes in the laboratory is an indication that proper quality control of production of concrete made with both fine and coarse recycled aggregate will be difficult if not impossible in practice. Similar observations were made by Mulheron /3/.

Table 1. Properties of natural materials and recycled aggregates

Type of Aggregate	Size Fraction in mm	Specific Gravity	Water Absorption in percent
Original material	0- 4	2.53	2.40
	4- 8	2.50	1.09
	8-16	2.62	0.96
	16-32	2.61	0.59
Recycled H	0- 4	2.21	12.56
	4- 8	2.34	9.92
	8-16	2.44	6.36
	16-32	2.50	5.81
Recycled M	0- 4	2.25	13.05
	4- 8	2.39	10.05
	8-16	2.47	5.26
	16-32	2.48	4.13
Recycled L	0- 4	2.27	11.44
	4- 8	2.38	8.98
	8-16	2.40	6.14
	16-32	2.48	3.34

3.4 Concrete Mix Proportions

Mix proportions for one cubic meter of the three original concretes are presented in Table 2, assuming both fine and coarse aggregates to be in saturated surface-dry condition. Adjustments in batch weights were made to correct for the true water contents of sand, gravel, and aggregates as batched. Original high-strength concrete (H) was produced with an effective water-cement ratio of 0.40, medium-strength concrete (M) with an effective water-cement ratio of 0.70, and low-strength concrete was produced with an effective water-cement ratio of 1.20. All fresh concretes achieved slump values between 60 and 100 mm.

All recycled concretes were produced with basically the same mix proportions as the original concretes except that small adjustments in weight proportions of aggregates were made in order to compensate for differences in density between original aggregates and recycled aggregates. All recycled concretes required an increase of 23 liters of mixing water per cubic meter of concrete after correcting for water absorption of recycled aggregates, in order to achieve approximately the

Table 2. Mix proportions for 1 cubic meter of concrete assuming aggregates to be in saturated and surface-dry condition

Type of concrete	Mix Component					
	Cement kg/m^3	Water kg/m^3	Fine Aggregate kg/m^3	Coarse Aggregate kg/m^3	Effective w/c	Density kg/m^3
H	402	161	542	1264	0.40	2369
M	230	161	719	1225	0.70	2335
L	134	161	908	1110	1.20	2313
H/H	459	184	516	1048	0.40	2207
H/M	459	184	520	1055	0.40	2153
H/L	459	184	510	1029	0.40	2107
M/H	262	184	649	1058	0.70	2218
M/M	262	184	656	1067	0.70	2169
M/L	262	184	658	1073	0.70	2124
L/H	153	184	867	903	1.20	2182
L/M	153	184	876	911	1.20	2177
L/L	153	184	877	912	1.20	2126

same slump values as the original concretes. Cement contents were increased correspondingly to maintain the same effective water-cement ratios in recycled concretes as in original concretes. Mix proportions for one cubic meter of the original three and the nine recycled concretes are presented in Table 2.

3.5 Production of Concretes

One set of twenty 100 mm by 200 mm standard cylinders was cast from each grade of original concrete and cured in water at 20 C. The compressive strength for ten cylinders of each set was determined after 14 days and the remaining ten cylinders were tested after 204 days of curing in water at 20 C.

The remainder of each batch of original concrete was cast into a number of 150 mm by 300 mm cylinders for production of recycled aggregates. These cylinders were also cured for 204 days in water at 20 C. After 204 days the cylinders were split and passed through a laboratory jaw crusher. The crusher products were screened and recombined into recycled high-, medium-, and low-strength grades, all having approximately the same grading as the natural aggregates from which the original concretes were produced.

New high-strength, medium-strength, and low-strength concretes were then produced from each of the three grades of recycled aggregate. The recycled concretes were designated H/H, H/M, H/L, M/H, M/M, M/L, L/H, L/M, and L/L, where the first letter designates the grade of strength of the recycled concrete (high, medium, or low) and the second letter designates the grade of strength of the original concrete from which the recycled aggregates were derived (also high, medium, or low). The compressive strength of five 100 mm by 200 mm cylinders from each recycled concrete was tested after 14 days and the remaining five cylinders were tested after 204 days of curing in water at 20 C.

4. Experimental Results

Properties of original and recycled aggregates are presented in Table 1, while mean values of concrete compressive strength test results are presented in Table 3.

5. Discussion

5.1 Properties of Natural and Recycled Aggregates

It is apparent from results which are presented in Table 1 that compared to natural sand and gravel, recycled aggregates have lower density and higher water absorption because considerable quantities of old mortar remain attached to sand and gravel particles in recycled aggregates.

5.2 Properties of Recycled Concretes

The 23 l per cubic meter higher free water requirement of recycled concrete made with both coarse and fine recycled aggregate can be explained by the fact that the recycled aggregates consist of crushed and angular particles with rough surfaces while natural materials consist of rounded particles which have relatively smooth surfaces. However, it is interesting that in a parallel investigation of similar recycled concretes made with coarse recycled aggregate and natural sand

Table 3. Compressive strength in MPa of original and recycled concretes *made with both coarse and fine recycled aggregate*

Curing Time	H	H/H	H/M	H/L	M	M/H	M/M	M/L	L	L/H	L/M	L/L
14 days in water at 20 C	49.5	37.3	33.6	33.7	23.9	16.1	17.2	19.1	8.7	5.5	4.5	6.8
204 days in water at 20 C	56.1	51.4	45.7	38.9	38.9	24.9	25.8	24.3	17.0	9.3	6.8	10.3

Table 4. Compressive strength in MPa of original and recycled concretes *made with natural sand and coarse recycled aggregate* (from /2/)

Curing Time	H	H*/H*	H*/M*	H*/L*	M	M*/H*	M*/M*	M*/L*	L	L*/H*	L*/M*	L*/L*
14 days in water at 20 C	49.5	54.4	46.3	34.6	26.2	27.7	27.0	23.2	9.1	10.2	10.3	9.6
11 days in water at 20 C + 27 days in water at 40 C	56.4	61.4	49.3	34.6	34.4	35.1	33.0	26.9	13.8	14.8	14.5	13.4

Symbols H, M, and L indicate original high-strength, medium-strength, and low-strength concretes made with natural gravel and sand. Symbol H/M indicates a high-strength recycled concrete *made with both fine and coarse recycled aggregate* produced from original medium-strength concrete, etc. Symbol H*/M* indicates a high-strength recycled concrete *made with natural sand and coarse recycled aggregate* produced from medium-strength concrete, etc.

which was carried out by Hansen and Narud /1/, the free water requirement was only 10 l per cubic meter higher for recycled than for original concretes. Thus recycled fine aggregates increase the water requirement of concrete much more than do recycled coarse aggregates.

The lower density of recycled concretes compared with the density of original concretes is due to considerable amounts of old mortar of relatively low density which remain attached to original sand and gravel particles in recycled aggregates. As recycled aggregates have lower density than natural aggregates, recycled concretes also have lower density than original concretes.

5.3 Compressive Strength of Hardened Concretes

It may be concluded from the data presented in Table 3 that on average 14-day compressive strengths of the nine recycled concretes made with both coarse and fine recycled aggregates were 27 percent lower than the 14-day compressive strength of corresponding original concretes made with similar free water-cement ratios. After 204 days of curing in water at 20 C the average reduction in strength was 34 percent.

In a parallel investigation which was carried out by Hansen and Narud /1/ of similar concretes where coarse recycled aggregates were used with natural sand, there was on average no difference in 14-day compressive strength between recycled concretes and control concretes, and the average reduction in strength of recycled concretes compared with control concretes after an additional 38 days of accelerated curing at 40 C was only 5 percent (see Table 4). Only when Hansen and Narud attempted to produce high-strength recycled aggregate concretes from original lower-strength concretes were higher strength reductions observed.

Thus it appears that the use of fine recycled aggregate consistently reduces the compressive strength of recycled concrete much more than does coarse recycled aggregate. Considering the large reduction in strength together with the fact that fine recycled aggregates also increase the water requirement of recycled concrete and that there is doubt about the frost resistance of such concrete /1/, it appears reasonable to conclude that recycled fine aggregates should not be used for production of quality concrete.

It has been mentioned before that ASTM C128 is not an accurate method of determining when recycled fine aggregate is in a saturated and surface-dry condition. Thus the desired effective water-cement ratios of the recycled concretes have probably not been achieved with sufficient accuracy in this investigation to warrant a more detailed analysis of the data in Table 3.

6. Conclusions

a. It has been found that based on equal slump the water requirement of recycled concrete made with both *coarse and fine recycled aggregates* was 23 l per cubic meter of concrete or 14 percent higher than that of control concretes made with natural sand and gravel. In an earlier investigation it was found that the additional water requirement for similar recycled concretes made with *recycled coarse aggregates and natural fine aggregate* was only 10 l per cubic meter of concrete.

b. It has been found that based on equal water-cement ratios the use of both *coarse and fine recycled aggregates* on average reduces the compressive strength of recycled concretes by approximately 30% compared to control concretes made with natural sand and gravel. When recycled concretes are made with *coarse recycled aggregate and fine natural aggregate*, the strength loss is in most cases much smaller when compared to control concretes made with natural sand and gravel.

c. It has been found that the use of ASTM C128 "Standard Test Method for Specific Gravity and Absorption of Fine Aggregate" to assess when recycled fine aggregates are in a saturated and surface-dry condition is inappropriate and highly inaccurate. As a consequence it will be difficult to control the effective water-cement ratio of a concrete production whether in the laboratory, in a ready-mix plant or on site if concrete is produced with recycled fine aggregate.

d. Considering Items a, b, and c, it cannot be recommended to use recycled fine aggregate for production of quality concrete. This is a disappointing fact, but it should be remembered that crushed concrete fines can be used for many other useful purposes.

Literature References

/1/ Hansen, T.C.: "Recycled Aggregates and Recycled Aggregate Concrete - Second State-of-the-Art Report. Materials and Structures (RILEM), Vol. 19, No. 111, pp.201-204, May-June 1986.

/2/ Hansen, T.C., and Narud, H.: "Strength of Recycled Concrete Made from Crushed Concrete Coarse Aggregate". Concrete International - Design and Construction (ACI), Vol. 5, No. 1, pp.79-83, January 1983.

/3/ Mulheron, M.: "A Preliminary Study of Recycled Aggregates". Institute of Demolition Engineers (GB). November 1986, p.39.

EFFECT OF IMPURITIES IN RECYCLED COARSE AGGREGATE UPON A FEW PROPERTIES OF THE CONCRETE PRODUCED WITH IT

K.YANAGI Japan Testing Centre for Construction Materials
M.NAKAGAWA Toda Construction Co., Ltd.
M.HISAKA Japan Testing Centre for Construction Materials
Y.KASAI College of Industrial Technology, Nihon University

Abstract
This paper deals with the test results of recycled coarse aggregate for concrete sampled from two processing plants in the suburbs of Tokyo. The quantity and kinds of impurities in demoilition waste of concrete were examined and three washing methods for removing impurities were tested. The physical properties of recycled coarse aggregate, and the compressive strength, drying shrinkage and carbonation rate of washed recycle-coarse aggregate concrete were tested to make clear the relationship between the washing methods of impurities and the physical properties of concrete made from them.
<u>Key words</u>; Recycled coarse aggregate, Recycled coarse aggregate concrete, Quantity of impurities, Mix proportion, Compressive strength, Drying shrinkage, carbonation .

1.Introduction

The need has recently arose for the reuse of recycled aggregate such as demolition waste of concrete for building industry in Japan. In April 1974, "the Research Committee on Disposal and Reuse of construction Waste" was established by Building Contractors Society (BCS)in Japan, and prescribed "Proposed standard and for the use of recycled aggregate concrete and commentary" (1977). About 4 years later, the similar project team was started by Ministry of Construction to confirm the results of BCS. In this paper, about 20 samples of recycled aggregate from two different processing plants in the suburbs of Tokyo were tested.

The amount of impurities and fines in recycled aggregate were examined, and in this test a specification was made for washing method of recycled aggregate to remove impurities practically. The physical properties of tests were conducted for the washed recycling coarse aggregate and crushed stone. The compressive strength, drying shrinkage and carbonation test were carried out for recycled aggregate concrete and crushed stone aggregate concrete for a comparison. The impurities in recycled aggregate were subjected to the chemical analysis, X-ray diffraction and DTA analysis.

2. Sampling of recycled aggregate

The following three series of test were conducted using recycled coarse aggregate of 20 samples collected from Plants N and S. These materials were sampling for three months in 1987. Plant N produced recycled aggregate from demoiltion waste of reinforced concrete (RC) structures, while Plant S was producing recycled aggregate from demolition waste of various structures including RC structure.

2.1 Experimental series

The experiments were composed of three series shown as belows;

Series I: Experiment on compressive strength, drying shrinkage and carbonation test of recycled aggregate concrete.

Series II: Experiment on impurities and fine particles contained in the recycled coarse aggregate.

Series III: Experiment on chemical analysis, X-ray diffraction and differential thermal analysis and visual analysis of impurities in recycled aggregate.

3. Experimental series I

3.1 Materials Ordinary portland cement was used. Physical properties of crushed stone coarse aggregate, fine aggregate and recycled coarse aggregate are shown in Table 1. Air entraining agent was used. Mixing water was ion exchanged pure water.

Table 1 Physical properties of recycled coarse aggregate, crushed stone coarse aggregate and fine aggregate

Mark		Number	Specific gravity Saturated surface dry	Specific gravity Oven dry	Absorption %	Bulk density of aggregate Kg/l	Loss in washing test %	Cement paste adhering to the old aggregate %	Finess modulus F.M.	Amount of light particles %	Sampling date
Crushed stone aggregate			2.65	2.63	0.58	1.63	0.4	–	6.78	–	–
Fine aggregate			2.64	2.60	1.84	1.72	1.4	–	2.76	–	–
Recycled coarse aggregate	N	1	2.48	2.37	4.72	1.40	2.68	15.1	6.48	5.02	June 16,87'
		2	2.46	2.33	5.24	1.35	2.12	12.4	6.55	3.48	23,
		3	2.46	2.35	4.98	1.33	1.96	13.2	6.51	4.22	30,
		4	2.46	2.34	4.80	1.37	1.76	11.0	6.51	0.68	July 21,
		5	2.46	2.34	4.73	1.38	2.13	12.9	6.34	0.90	28,
		6	2.47	2.36	4.52	1.52	1.38	11.5	6.42	0.55	Aug 11,
		7	2.50	2.39	4.45	1.54	0.91	12.2	6.57	4.99	18,
		8	2.44	2.32	5.28	1.33	1.06	13.8	6.53	2.24	25,
		9	2.44	2.32	5.24	1.36	1.57	13.5	6.57	1.99	Sept 1,
		10	2.44	2.34	4.76	1.48	1.13	12.4	6.34	3.86	8,
		$\bar{X}$	2.46	2.35	4.87	1.40	1.67	12.8	6.48	2.79	
		σ	0.12	0.02	0.30	0.08	0.56	1.2	0.09	1.75	
		C.V(%)	0.8	0.9	6.2	5.7	33.5	9.4	1.4	62.7	–
Recycled coarse aggregate	S	1	2.32	2.14	8.46	1.26	3.09	23.6	6.78	13.10	June 11,
		2	2.34	2.18	7.76	1.18	1.18	19.6	6.50	8.31	18,
		3	2.35	2.18	7.68	1.34	2.26	18.3	6.24	4.96	25,
		4	2.36	2.20	7.59	1.34	3.06	19.4	6.28	7.79	July 9,
		5	2.26	2.07	9.70	1.22	3.05	15.0	6.45	15.88	16,
		6	2.46	2.36	4.49	1.30	2.09	22.2	6.49	8.23	23,
		7	2.36	2.19	7.62	1.31	2.87	19.9	6.35	4.49	30,
		8	2.32	2.14	8.24	1.28	3.40	19.9	6.45	5.60	Aug 6,
		9	2.36	2.19	7.58	1.32	2.82	19.4	6.28	4.87	13,
		10	2.42	2.28	5.81	1.42	2.01	19.5	6.54	2.24	Sept 3,
		$\bar{X}$	2.36	2.19	7.49	1.30	2.58	19.7	6.44	7.55	
		σ	0.06	0.08	1.43	0.07	0.68	2.3	0.16	4.17	
		C.V(%)	2.5	3.7	19.1	5.4	26.4	11.7	2.5	55.2	–

3.2 Experimental methods

(1) Mix proportion of concrete and making of concrete specimen
Mix proportion of concrete was; water-cement ratio: 0.6, sand-aggregate ratio: 0.44, slump: 18 cm and air content: 4 %. Concrete specimens were made by the method of JIS A 1132. The size of specimen was 10 x ϕ 20 cm cylindrical specimen.

(2) Compressive strength The specimen was cured in water at 20 °C and in air at 20 °C and 60 % RH. The compressive strength of concrete was tested in accordance with JIS A 1108 at the age of 28 days.

(3) Drying shrinkage The drying shrinkage of concrete was tested by JIS A 1129.

(4) Carbonation Size of specimens for accelerated carbonation test was 10 x 10 x 40 cm in rectangular prism. Specimens were demoulded after curing for 2 days in a 20 °C and 70 % RH room. Thereafter they were cured for 28 days in air of 20 °C and 60 % RH. The specimens were kept in an accelerative carbonizing test room of 20 °C, 60 % RH and 5 % carbon dioxide. The depth of carbonation was 10 cm or more after the specimens were kept for 1 - 2 months in the carbonizing room mentioned above. The carbonation depth was measured by 1% phenolphthalein-ethanol solution.

3.3 Results and Discussions

(1) Quantity of impurities The quantity of impurities contained in recycled coarse aggregate was varied widely with sampling date and producing Plants of N and S (refer to Table 1). Each producing plant has a peculiar acceptance standard of demolished waste concrete and this difference of standard gives great influence on the quality of recycled aggregate.

When a specific gravity become larger the quantiy of impurities tends to become smaller, but correlation is not so high (refer to Fig.1).

(2) Mix proportion of recycled coarse aggregate concrete If the particle size of aggregate is appropriate and conform to the standard

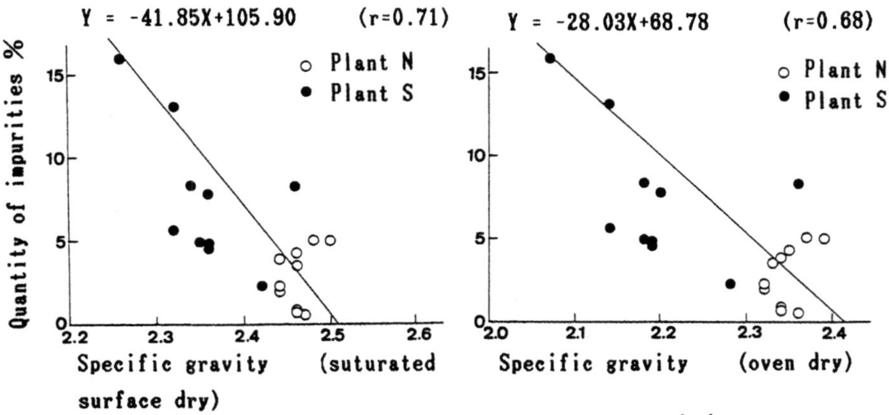

Fig. 1. Relation between specific gravity of recycled coarse coarse aggregate and quantity of impurities

grading range of JIS A 5308, the required value of slump of recycled coarse aggregate concrete can be obtained regardless of the quantity of impurities, except a few number of cases. If the correction of coefficient of aggregate is revised, the amount of AE agent obtaining a required quantity of air will be nearly equal to that of crushed stone concrete.

(3) Compressive strength Fig. 2 shows the compressive strength of recycled coarse aggregate concretes using non-washed aggregate. The compresive strength of them shows within 70 to 90 % of crushed stone concrete, but there is a difference between two producing Plants N and S. The strength of concrete using aggregate from Plant S seems to be smaller than the results of past studies BCS (1977) and Nakagawa et.al. (1987). In this case, the effect of amounts of impurities on compressive strength of recycled aggregate concrete is not so large, though the composition of impurities in raw concrete will be considered to play a dominant influence (Fig.2).

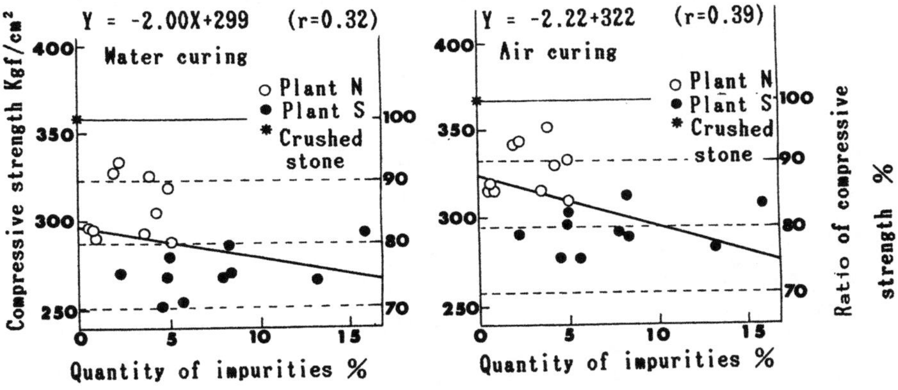

Fig. 2. Relation between quantity of impurities and compressive strength

The report of the BCS Committee (1977) gave indication of serious detrimental effects on the compressive strength of recycled aggregate concrete by impurities.
Decreasing rate of strength by impurities is in order as shown belows; PVAc paint > Asphalt > Gypsum hydrate > Wood > Soil > Plaster.
After the results of BCS committee, each maximum permissible impurities of aggregate is given about 85 % to the compressive strength of controled concrete(Table 2).

Table 2 The maximum permissible value of impurities corresponding to concrete

Impurities	plaster	soil	wood Japanese cypress	gypsum hydrated	asphalt	vinyl acetate paint
Volume %	6	5	4	3	1	0.2

(4) Drying shrinkage Where as drying shrinkage of recycled aggregate concrete using aggregate from Plant N is 7.64 to 8.80 $\times 10^{-4}$ (average 8.34 $\times 10^{-4}$) , that of concrete using aggregate from Plant S is 9.14 to 10.76 $\times 10^{-4}$ (average 9.96 $\times$ 10^{-4}). Fig.3 shows the effect of the quantities of impurities of concrete using non-washed aggregate on the drying shrinkage. The value of the drying shrinkage of concrete using aggregate from Plant N is large to 129 % in comparison with that of curshed stone concrete and this result is agreed with that of Kasai(1976) but concrete using aggregate from Plant S takes 153 % which is greater than the result of BCS (1977).

(5) Carbonation The value of carbonation is generally higher in concrete using aggregate from Plant S than in concrete with aggregate from Plant N (Fig. 4).

Carbonation rate of concrete using aggregate from Plant N gives 1.28 times in comparison with crushed stone concrete. This value nearly equal to the value of BCS (1977), while the value of concrete using aggregate from Plant S is 1.53 times in the average and greater than the results of BCS (1977).

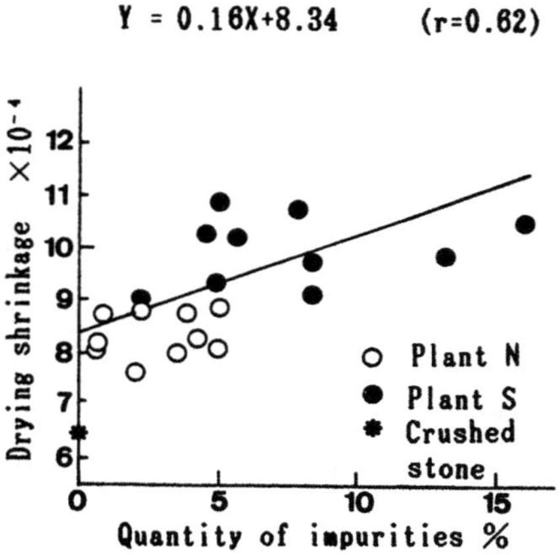

Fig. 3. Relation between drying shrinkage and quantity of impurities

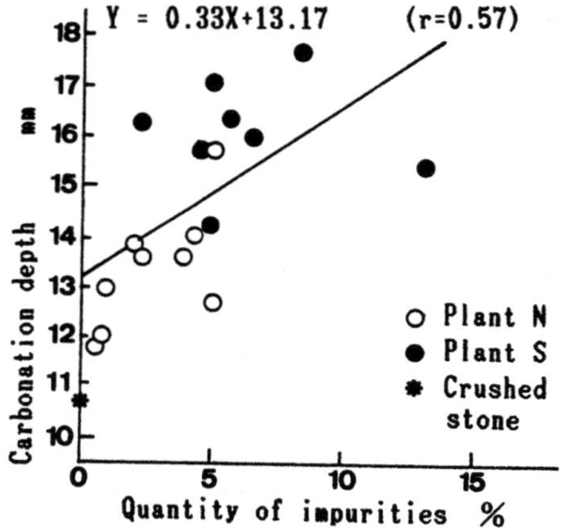

Fig. 4. Relation between carbonation depth and quantity of impurities

4. Experimental Series II

4.1 Sampling for test Six samples were tested in this serise and . were N-1, -2, -3 and S-1, -2, -5 as the number of shown in Table 1.

4.2 Method of experiment

(1) Removal processes of impurities and fines from recycled aggregate

(a) Washing method I (mark:W): The sample of coarse aggregate of 10 kg was thrown into a container filled with 60 litre water was supplied into the container continuously at the rate of 0.4 litre per minute. The coarse aggregate was kept twice for one minute each (Fig.5).

(b) Washing method Ⅱ(mark:SW10): An electromotive sieve was used and washed by continous watering shower of 20 litre per minute, 10 kg sample of coarse aggregate was washed for 10 seconds (Fig. 6).

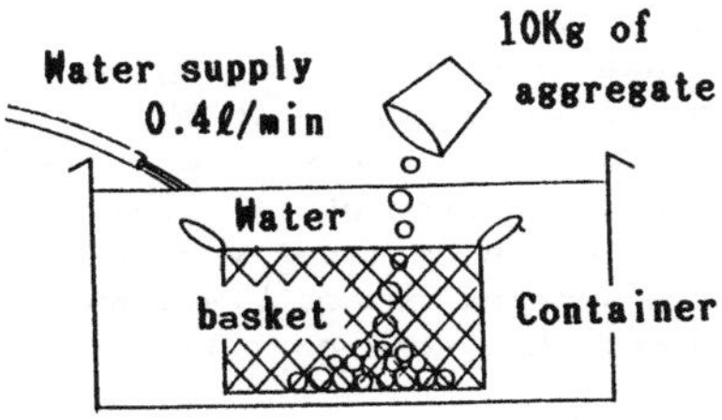

Fig. 5. Washing method Ⅰ (Mark: W)

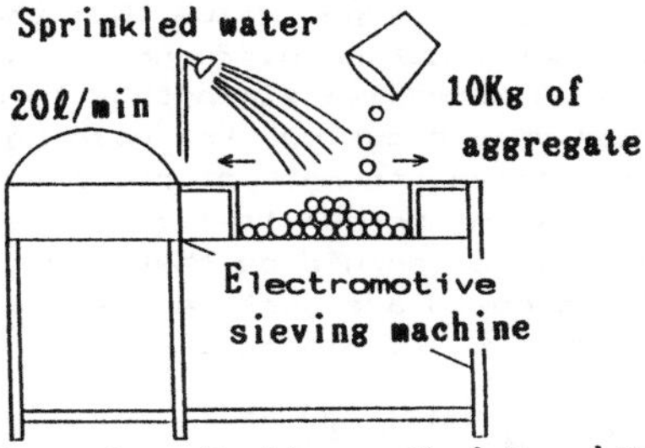

Fig. 6. Washing method Ⅱ and Ⅲ (Mark:SW10,SW 60)

(c) Washing method Ⅲ(mark:SW60): Washing was operated for 60 seconds as the method mentioned in washing method Ⅱ.

(2) Quality test of recycled aggregate after washing Loss of washing was measured and quantity of impurities was tested based on JIS A 1103 testing method.

(3) Physical properties of concrete using washed recycle-aggregate The physical properties were tested by the same methods of experimental series Ⅰ.

4.3 Results and Discussions

(1) Effect of the method of washing on loss of weight and quantity of impurities of recycled coarse aggregate Fig.7,8 shows the method of loss of washing and quantity of impurities, correspondence to difference of the washing method. The recycled coarse aggregate from Plant N containes large amount of fines and impurities than that of Plant S. The order of loss of washing is shown as belows: non-washing (N) < washing method Ⅰ(W) < washing method Ⅱ(SW10)< washing method Ⅲ(SW60)

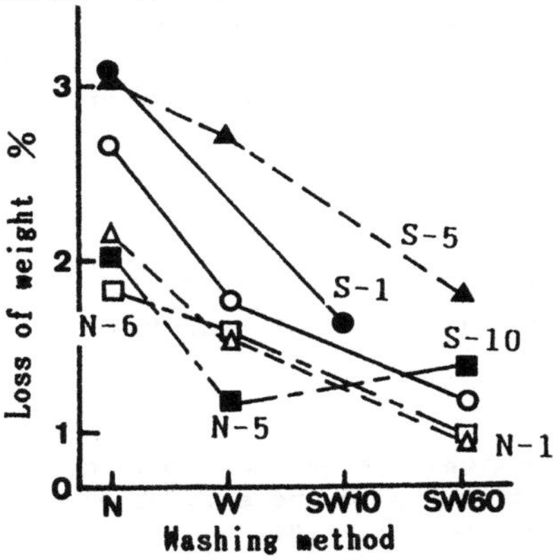

Fig. 7. Effect of the method of washing on loss of weight

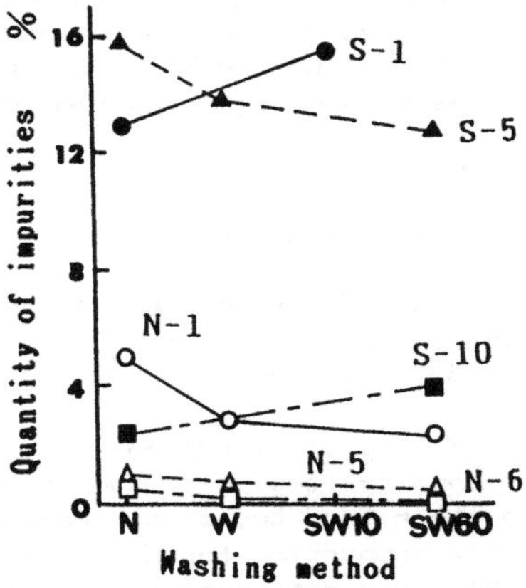

Fig. 8. Effect of the method of washing on quantity of impurities

(2) Mix propotion of concrete using washed recycle-coarse aggregate The slump of concrete using washed recycling coarse aggregate shows a tendency of larger value than that of concrete using non-washed aggregate, but no differnce in amount of air.

(3) Compressive strength In either case of standard curing in water or air-curing, the washing processing is effective on the compressive strength of recycled coarse aggregate concrete, especially the washing method I(W) gives large compressive strength (Fig.9). This reason should be that the fine particles which cover the surface of aggregate are just removed away by the washing method I(W), and appropriate fine particles are remained and they work as filler.

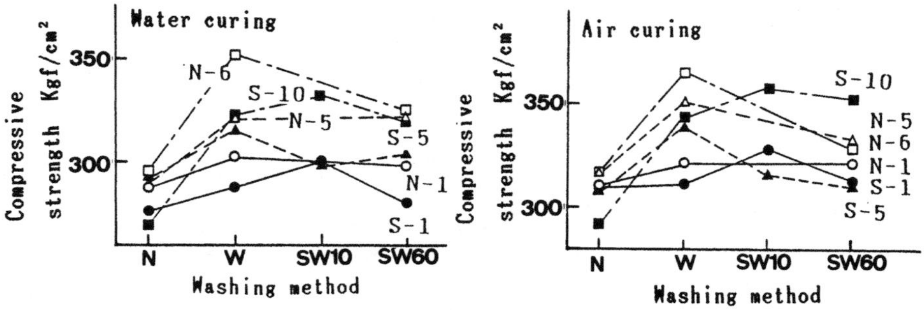

Fig. 9. Relation between washing method and the compressive strength

(4) Drying shrinkage Drying shrinkage of concrete using recycled aggregate shows in Fig.10. The shrinkage of concrete using the aggregate from Plant S larger than that of Plant N. Because the quality of aggregate from Plant S is lesser than that of Plant N. Drying shrinkage of concrete using recycled aggregate at the age of 13 weeks shows large scatter depending upon washing method.

(5) Carbonation When one month of accelerated test is completed the carbonation depth of washed recycle-aggregate concrete is larger than that of non-washed aggregate concrete. The reason should be that the washed aggregate was removed its fine particles by washing prosess (Fig.11).

5. Experimental series III

5.1 Plan of physicochemical analysis

Table 3 shows the plan of physicochemical anaysis of recycled aggregate concrete.

5.2 Method of Experiment

(1) Chemical analysis Some of the chemical analysis of recycled aggregate from two plants were made according to the estimation method of mix proportion for hardened concrete (CAJ ; F-18).

(2) X-ray diffraction analysis The equipment for this test was a rotary cathode type Geigerflex RAD-γA made by Rigaku denki Co. Ltd. Material for analysis was crushed under 0.6 mm sieve, thereafter dried and finally atmomized into micro particle.

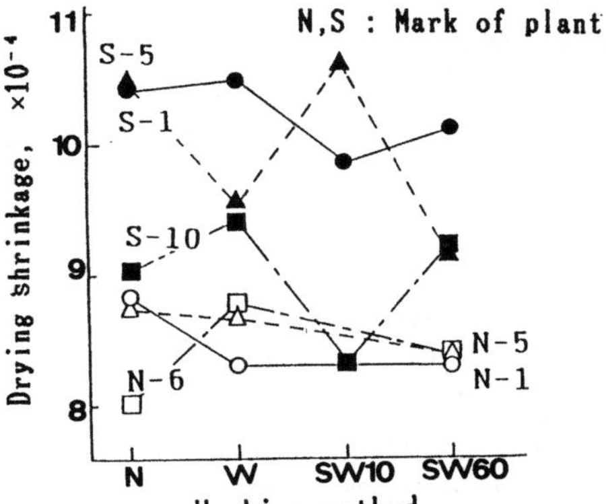

Fig.10. Relation between washing method of aggregate and drying shrinkage of concrete

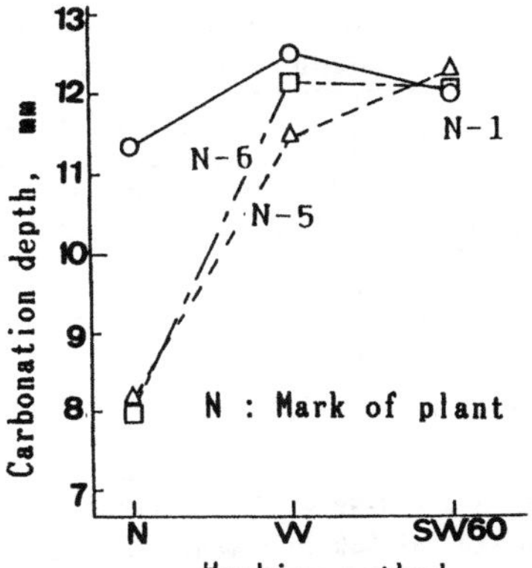

Fig.11. Relation between washing method of aggregate and carbonation depth of concrete

Table 3 Plan of physicochemical analysis

Sample [1]	Chemical analysis	X-ray diffraction analysis	TG & DTA [2]
No.1	○	○	－
No.2	－	○	○
No.3	－	○	○

Note,

1) No.1 : Recycled aggregate under 5mm sieve.
No.2 : Impurities in recycled aggregate which float in zinc chloride solution of 1.95 in specific gravity.
No.3 : Recycled coarse aggregate washed with water.
2) TG & DTA : Thermal gravity and differential thermal analysis.

(3) Differential thermal analysis (DTA) High temperature type TG (thermal gravity)- DTA was used and a material for DTA was the same material as X-ray diffraction analysis.

(4) Visual inspection The composition of impurities in recycled aggregate from two plants were investigated by a visual inspection.

5.3 Results and Discussions

(1) Chemical anaiysis Quantity of cement of aggregate from Plant S obtained from the anlysis of mix proportion of concrete is larger than that of Plant N (Table 4).

(2) X-ray deffraction analysis The result of X-ray diffraction anaiysis is shown in Table 5. The particles (impurties) which float in the heavey liquid having a specific gravity of 1.95 containing monosulfate and Friedel's salt, and also calcium hydroxide is characteristically found in washed aggregate. The chloride content shows a slight difference, but no particular trace of inclusion of foreign matters is found in either plants.

Table 4 The result of chemical analysis of recycled aggregate and composition of recycled concrete

Item / Subject	Chemical analysis %				Composition of recycled concrete	
	ig . loss (600℃)	ig . loss (1000℃)	insol	CaO	cement-wt estimated	Aggregate** -wt estimated
Plant N	4.5	8.8	72.5	9.4 (12.8)*	15.4	84.4
Plant S	4.9	8.2	76.2	10.6 (19.7)*	16.6	83.4

Note,*():Calcium oxide is solved by 0.5 percent hydrochloride solution.(percent.)
**Content of CaO in cement : 64.5 percent. CaO-content in aggregate : 0.4percent.
Insol ratio in aggregate : 95.2 percent.
Both ratio CaO and aggregate : 100 percent.

Table 5 Result of X-ray diffraction

Minerals / Sample		Quartz	Feldspar	Mica	Chlorite	Hornblende	Calcite	Mono sulfate	Flieder's salt	Dolomite	Calcium hydroxide
	No.1	v	iv	i	i	ii	v	i	i	ii	i
Plant N	No.2	v	iv	i	i	i	iii	i	i	i	i
	No.3	v	iv	i	i	ii	iii	i	i	ii	ii
	No.1	v	iv	i	i	ii	v	i	i	ii	i
Plant S	No.2	v	iv	i	i	i	iii	i	i	i	i
	No.3	v	iv	i	i	i	i	i	i	ii	ii

Note, i ; trace. ii ; little. iii ; medium. iv ; much. v ;very much.

(3) Differential thermal analysis DTA curve of recycled aggregate is not much different from that of crushed stone concrete, but in the case of impurities alone, the continuous loss in weight is observed as indicating the difficulty in anaiysis.

(4) Visual investigation Content of impurities in recycled aggregate differ from Plant N and S. The aggregate from Plant S contains much impurities such as pumice and lightweight aggregate mortar than that of Plant N (Table 6).

6. Conclusion

(1) Quantities of impurities contained in recycled coarse aggregate in the 20 samples collected from two plants are within the range of 0.7 % to 15.9 %, and a significant difference is observed between the two plants. This is considered that there is a clear difference in quantity of raw concrete rubble themselves which were received at the either plants.

Table 6 Impurities visually inspected in recycled aggregates

Subject of Test	Impurities in Agrregate	
	Inorganic	Organic
Plant N	Gypsum plaster. Light weight concrete. Coal cinder concrete. Pumice stone. Pumice mortar.	Asphalt. Piece of plastics. Piece of wood. Vinyl tube. Paper. Piece of cloth.
Plant S	Light weight concrete. Pumice mortar. Pumice stone. Gypsum plaster. Piece of Ooya stone. Cement paste.	Asphalt. Piece of wood. Paper. Vinyl tube.

(2) There is a relationship between the quantity of impurities and the compressive strength, drying shrinkage and carbonation of concrete made from recycled coarse aggregate, and the physical properties of concrete using recycled aggregate from two plans are obviously different because of the difference of acceptance standard of raw concrete rabble.

(3) When recycled coarse aggregate is washed with water in three grade;

The compressive strength of recycled coarse aggregate concrete is affected by washing method. The washing method I (W) gives large compressive strength.

There is no significant difference in the drying shrinkage of recycled aggregate concrete caused by washing method.

Carbonation depth of washed recycle-aggregate concrete is larger than that of non-washed recycle aggregate concrete.

(4) The constitution of impurities found by visual inspection shows good agreement with the results of the chemical analysis and X-ray diffraction. Thus it can be said that the effect of impurities, which cannot be detected by X-ray diffractions, on physical properties of concrete can be ignored.

References

Building Contractors Society (1977), "The Recommended Practice for Usage of Recycled Aggregate and Recycled Concrete".

Building Reserch Institute Ministry of Construction (1986), "Report on Studies Concerning Recycling Technologies for Waste in Construction Project".

KASAI, Y.(1976), "Usage of Recycled Aggregate and Recycled Concrete" Concrete Journal, JCI Vol. 14, No. 9, Sept.

NAKAGAWA, M., YANAGI, K., HISAKA, M., KEMI, T.,(1987), "Fundamental Investigtion of impruities in Concrete Contaminated by Recycled Aggregate" Annual Meeting, AIJ.

DURABILITY OF CONCRETE USING RECYCLED COARSE AGGREGATE

Y. KASAI
College of Industrial Technology, Nihon University, Japan

M. HISAKA AND K. YANAGI
Japan Testing Centre for Construction Materials, Japan

Abstract
Durability is one of the most important characteristics of recycled aggregate concrete. The compressive strength, drying shrinkage, accelerated carbonation, resistance to freezing and thawing, and water absorption under pressure are tested using recycled coarse aggregate sampled from two plants.

An ordinary coarse aggregate (crushed stone aggregate) was replaced with ratios of 0, 30, 50 and 100 percent by recycled coarse aggregate from two plants and mixed with water-cement ratios of 50, 60 and 70 percent and tested for above mentioned properties
Key words : Recycled aggregate concrete, Compressive strength, Drying shrinkage, Accelerated carbonation, Resistance to freezing and thawing, Water absorption under pressure

1. Introduction

Important points for practical use of recycled aggregate concrete include mix proportions, strength, drying shrinkage, carbonation rate, resistance to freezing and thawing, air and water permeability, etc. in comparison to ordinary aggregate concrete.

Some of these properties were reported by the Building Contractors Society of Japan (BCS)(1975-1977). This paper deals with an assessment of recycled coarse aggregate sampled from two plants in terms of four replacement ratios of recycled coarse aggregate to ordinary coarse aggregate, three water-cement ratios and two curing methods for concrete, evaluating mix proportions, strength, drying shrinkage, carbonation and resistance to freezing and thawing on the basis of experiments. The test results of Kasai (1975) are introduced for water absorption under pressure.

2. Experimental method

2.1 Materials

(a) Coarse aggregate; Recycled coarse aggregate was sampled from two plants N and S which are described by the authors (1988) in another paper of this Second RILEM DRC Symposium.

The recycled coarse aggregate was defined by the size passing through the 20 mm sieve and remaining on the 5 mm sieve. Coarse aggregate for the controlled concrete used crushed stone aggregate : passing the 20 mm sieve and remaining on the 5 mm sieve.

Physical properties of the coarse aggegates are shown in Table 1. Table 2 indicates grain size distribution. Recycled concrete was made on crushed stone aggregate replaced by 0, 30, 50 and 100 percent of recycled coarse aggregate for studying the effect of these replacement ratios on various properties of concrete.

(b) Fine aggregate Fine aggregate was river sand. The physical properties and grain size distribution are shown in Table 1 and Table 2 respectively.

(c) Other materials; Cement : ordinary portland cement, Admixture : AE agent (Vinsol), Water : tap water treated with an ion exchange resin.

2.2 Experimental method

Table 3 shows the water-cement ratios of concretes, replacement ratios of recycled aggregate, properties tested and other items.

Table 1 Properties of aggregates

Test item	Coarse aggrtegate C	Coarse aggrtegate N	Coarse aggrtegate S	Fine aggregate
Specific gravity of saturated surface dry	2.65	2.45	2.37	2.64
Specific gravity of absolute dry	2.63	2.31	2.23	2.60
Absorption of water(%)	0.58	5.63	6.50	1.84
Unit weight (kg/l)	1.63	1.37	1.32	1.72
Solid volume percentage for shape determination(%)	61.2	-	-	-
Quantity of clay lumps (%)	0.3	-	-	0.4
Loss of washing analysis of aggregate (%)	0.4	1.25	1.66	1.4
Organic impurlities test		-	-	Good
Loss of soundness test of aggregate (%)	7.1	-	-	5.3

Table 2 Grading of aggregates

Size of sieve mm		Percentage passing (%) 25	20	15	10	5	2.5	1.2	0.6	0.3	0.15	F.M.
Coarse aggregate	C	100	93	63	26	3	0	-	-	-	-	6.78
Coarse aggregate	N	100	100	94	53	3	0	-	-	-	-	6.44
Coarse aggregate	S	100	100	84	28	2	0	-	-	-	-	6.70
Fine aggregate		-	-	-	-	100	90	68	43	19	4	2.76

(a) Composition of concrete; Water-cement ratios of concete were 50, 60 and 70 percent using the recycled coarse aggregate from plant N. The aggregate from plant S was tested by 60 percent only. The slump was 18 cm and the air content was 4 percent respectively.

(b) Curing of concrte; The concrete was cured initially for one week in the water of 20 °C, then moved in air for three weeks at 20 °C, in general.

(c) Properties tested; Properties tested were compressive strength, drying shrinkage, accelerated carbonation rate and resistance to freezing and thawing. Detailed methods of these tests are described in another paper of this Second RILEM DRC Symposium by the authors(1988).

Table 3 Design of experiments

Testing items	Plant	W/C %	Slump cm	Air %	Replacement of recy.agg.%	Curing condition
Compressive strength	N S	50,60,70 60				(1),(2)
Drying shrinkage	N S	50,60,70 60	18	4	0,30,50,100	(3)
Carbonation depth	N S	50,60,70 60				(4)
Freezing & thawing	N S	50,60,70 60				(5)

(1) Standard curing (in water of 20 C) for 4 weeks and tested for compressive strength.

(2) Standard curing for 1 week, thereafter cured in the room of 20 C, R.H.60 % untill the age of 4 weeks and tested for compressive strength.

(3) Standard curing for 1 week, thereafter started the shrinkage test.

(4) Standard curing for 1 week, thereafter cured in the room of 20 C, R.H. 60 % untill the age of 4 weeks, thereafter started accelerated carbonation test.

(5) Standard curing for 2 weeks thereafter started the freezing and thawing test of 300 cycles.

3. Results and discussions

3.1 Compressive strength

(a) Effect of the replacement ratios and water-cement ratios on the compressive strength of recycled coarse aggregate; Trends of the compressive strength of concretes are shown in Fig.1 with the varied water-cement ratios and the replacement ratios of recycled coarse aggregate. Concrete with a smaller water-cement ratio provides smaller reduction of strength with increased replacement ratio of coarse aggregate. There is no significant difference between plants N and S, when the recycled aggregate concrete is made by water-cement ratio of 60 percent.

Referring to Fig.1 (a) and (b), the strength of concrete cured in water for 4-weeks is generally smaller. This is because, in case of concrete cured for 1-week in water thereafter cured in air for 3- weeks, sufficient moisture is supplied to concrete at the beginning period thereafter increase the strength by drying of 3-weeks period.

(b) Result of BCS on strength; Referring to the result of BCS (1978) when the recycled coarse aggregate was replaced by 100 percent the strength was reduced by about 15 percent. The reduction ratios of strength became larger when both the recycled coarse aggregate and recycled fine aggregate were used together, show about 40 percent reduction when 100 percent was replaced by them.

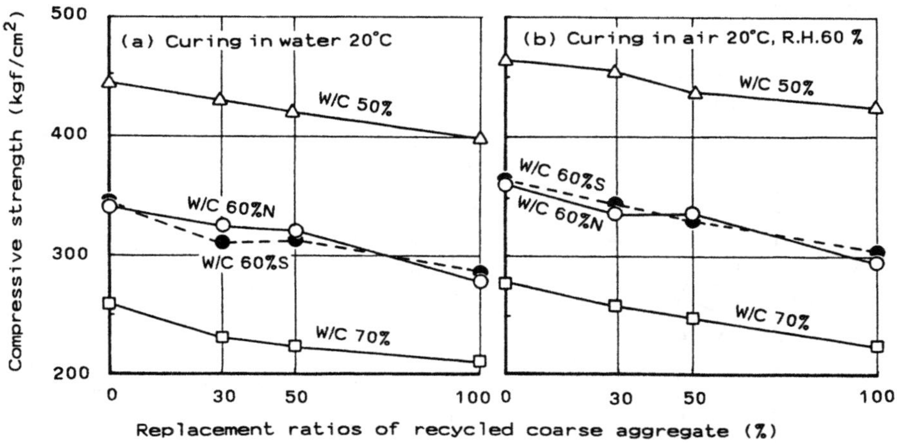

Fig.1 Replacement ratios of recycled coarse aggregate vs. compressive strength

3.2 Drying shrinkage

Relationships between the replacement ratio of recycled coarse aggregate and drying shrinkage at 4-weeks are shown in Fig. 2. The drying shrinkage of the recycled aggregate concrete is smaller than that of the controlled concrete for replacement ratios of 30 and 50 percent, though this reason is not obvious.

The drying shrinkage of concrete with a replacement ratio of 100 percent is equivalent to or slightly larger than that of the controlled concrete. According to the results of BCS (1977), the drying shrinkage ratios were 1.1 - 1.4 times as large at 4-weeks. In the present test, the drying shrinkage was generally smaller, but some data of indicated values were close to the upper limit of the BCS results. No particular relatioships were observed between the water-cement ratios and shrinkage. It is not obvious whether there is any difference between recycled coarse aggregates of plants N and S, because shrinkages of concrete were dispersed with recycled coarse aggregate.

3.3 Carbonation

Fig. 3 shows the results of the accelerated carbonation test by keeping the concrete in a vessel of carbonic dioxide gas of 5 percent concentration for 4-weeks. According to this result, depths of carbonation became larger as the water-cement ratios being larger.

However, no significant difference is observed even where the replacement ratios of recycled coarse aggregate are changed to 0, 30, 50 and 100 percent. These values are not much different from those of the controlled concrete. After the result of Yoda (1978), the carbonation depth of the recycled coarse aggregate concrete at 5-months was 1.2 times or more than that of the controlled concrete which was used river gravel and river sand.

From these results, it appears that the carbonation rate of recycled aggregate concrete will be the same or a little larger than that of ordinary aggregate concrete.

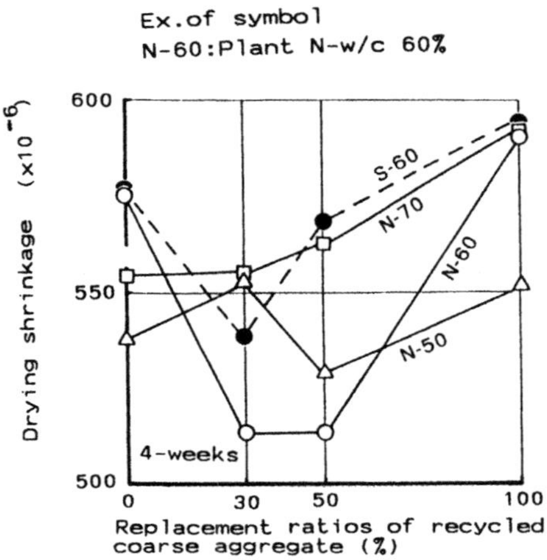

Fig.2 Replacement ratio of recycled coarse aggregate vs.drying shrinkage

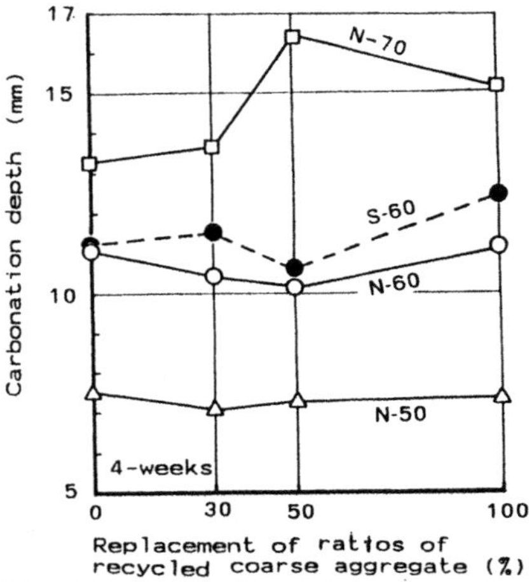

Fig.3 Replacement ratio of recycled coarse aggregate vs. accelerated carbonation depth

3.4 Resistance to freezing and thawing

Fig.4 shows the relationships between the number of freezing and thawing cycles and percentages of relative dynamic modulus of elasticity. Fig.5 shows the relative loss in weight of specimen crresponding to the specimens in Fig.4. Fig.6 shows the relationshis between the replacement ratio of recycled coarse aggregate and the durability factor (DF).

Fig.7 and 8 show the loss in weight of recycled aggregate concrete by the freezing and thawing cycles.

(a) Change of the relative dynamic modulus of elasticity and the relative loss in weight of recycled aggregate of concrete; The test for freezing and thawing cycles was started at the age of 2-weeks.

The modulus of elasticity of controlled concrete does not change significantly. It might be because the hydration of cement was still going on as the age was still so young that the cement was cured in water until the age of 2-weeks and then the test was begun.

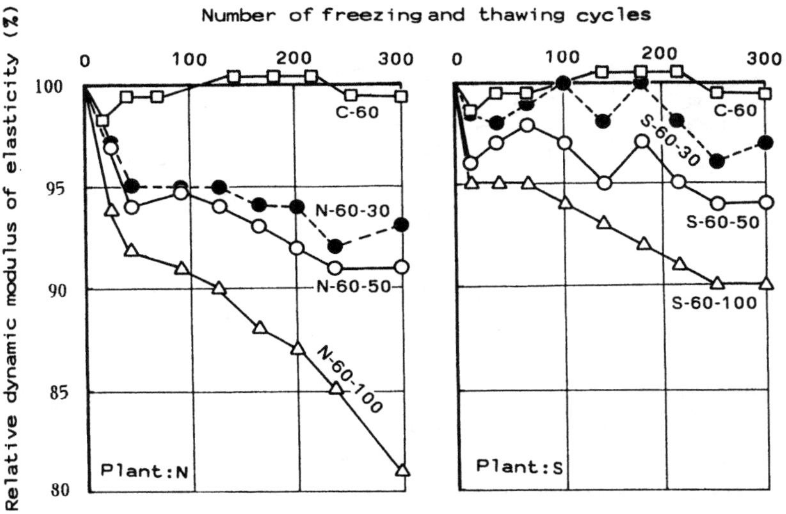

Fig.4 Number of freezing and thawing cycles vs. relative dynamic modulus of elasticity

Ex. of Marks

N-60-30

Plant N · W/C · Replacement ratios of r.c.a.

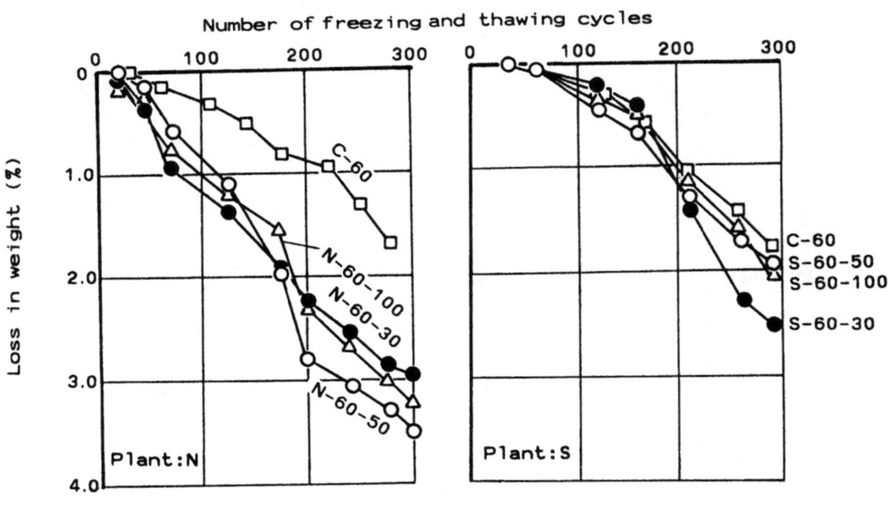

Fig.5 Number of freezing and thawing cycles vs. loss in weight

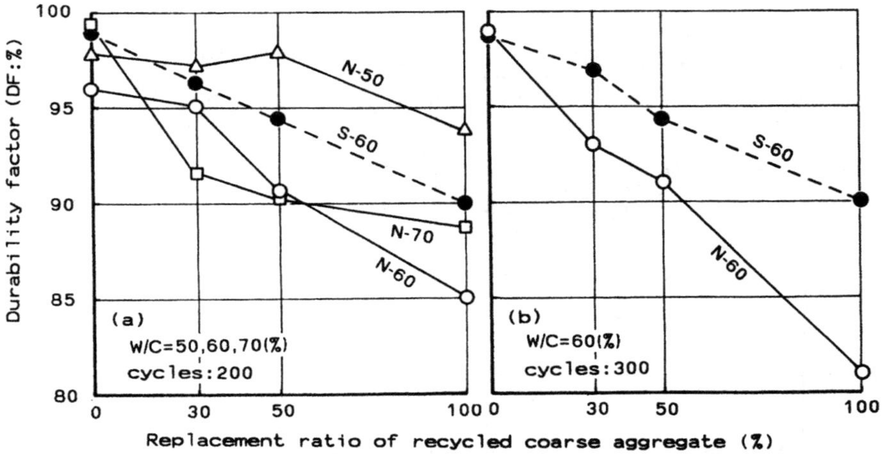

Fig.6 Replacement ratio of recycled coarse aggregate vs. durability factor

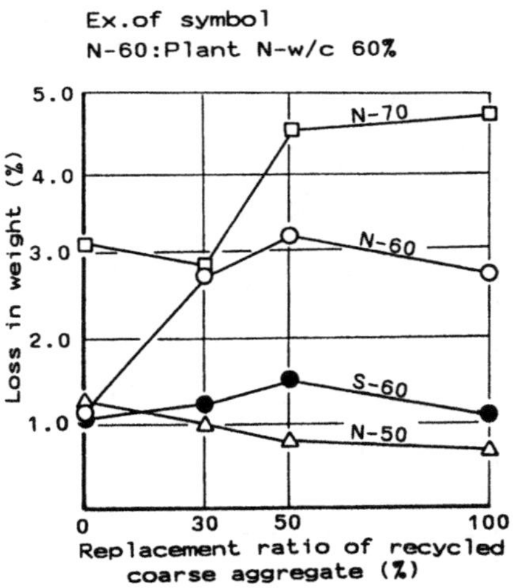

Fig.7 Replacement ratio of recycled coarse aggregate vs. loss in weight

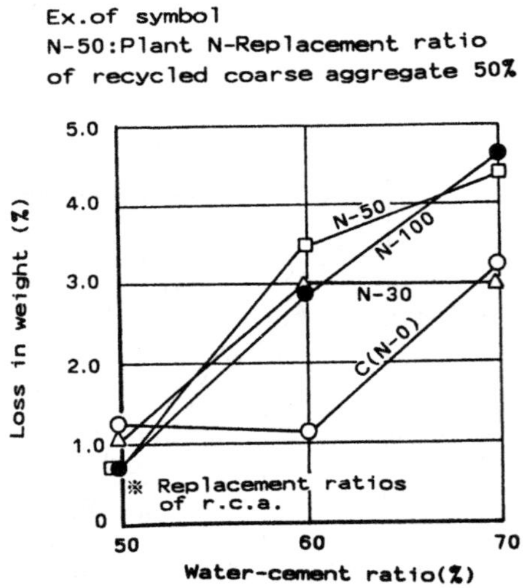

Fig.8 Water-cement ratio of recycled coarse aggregate concrete vs. loss in weight

Where the crushed stone aggregate was replaced by the recycled coarse aggregate, the relative dynamic modulus of elasticity and relative weight decrease as the numbers of freezing and thawing cycles increase.

(b) Relationships between the ratio of replacement of recycled coarse aggregate and the durability facter (DF); The DF value as shown in Fig.6 decreases as replacement ratio of recycled aggregate increases. Although decremental DF values are not very large, the freezing and thawing resistance of concrete becomes obviously smaller by replacing recycled coarse aggregate.

(c) Difference of durability factor between plants; Fig.6 shows the durability factor (DF) of concrete with 60 parcent in water-cement ratio which was made of recycled aggregate from plants N and S. The recycled coarse aggregate of plant N brought out larger decrement of DF value at 200 and 300 cycles. The aggegate from plant N provides larger specific gravity, less impurities of aggregate, and smaller absorption of water as shown in Table 1, namely better qualities than that of plant S though the strength of both concrete with the same water-cement ratios are nearly the same as shown in Fig.1 (b). However, the freezing and thawing resistance of aggregate from plant N becomes obviously inferior to that of plant S. The reason is not clear, but it is probably due to the air content in mortar and other composition in the recycled coarse aggregate of plant N.

(d) Test for resistance to freezing and thawing after BCS; According to the result of Y. Kasai described in the BCS report(1975), durability factor at 300 cycles for concrete using 100 percent of recycled coarse aggregate was smaller than that of the controlled concrete by 10 percent or less.

Consequently, it should be noted that the resistance of the recycled coarse aggregate concrete to freezing and thawing cycle is slightly inferior to that of the controled concrete.

4. Water absorption test of concrete under pressure

According to Y. Kasai discribed in the BCS report(1975), a water-absorption test is carried out under pressure using cylindrical test specimens, as outlined in the following.

4.1 Test specimens

(a) Recycled coarse aggregate; Composition of original concrete of recycled coarse aggregate was river gravel and sand with water-cement ratios of 45, 55 and 68 percent. These concrete were crushed with a jaw crusher. Sizes of recycled coarse aggregate were 25 mm or less.

(b) Mix proportions; Water-cement ratios of test concrete were 50, 60 and 70 percent with a slump of 21 cm. Kinds of concrete including a controlled concrete using river gravel and river sand are expressed as (NS, NG) and the recycled aggregate concrete with three of water-cement ratios using river sand and all recycled coarse aggregate is expressed as (NS, CG).

(c) Shape and dimensions; Shape and dimensions of cylindrical concrete test specimen were 150 mm in outer diameter, 19 mm in inner diameter and 300 mm in hight.

4.2 Water absorption test under pressure

The cylindrical test specimens, mentioned above were attached tố a pressure vessel using packings of the upper and lower parts, subjected to a water pressure (12kgf/cm²) externally for six hours while measuring the increase in weight of the test sample for evaluating water permeability in terms of water absorption rate per unit volume of concrete in gram per litre.

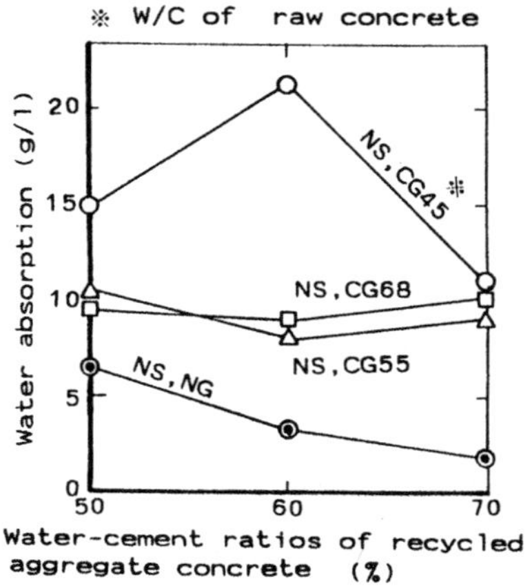

Fig.9 Water-cement ratios of recycled aggregate concrete vs. water-absorption

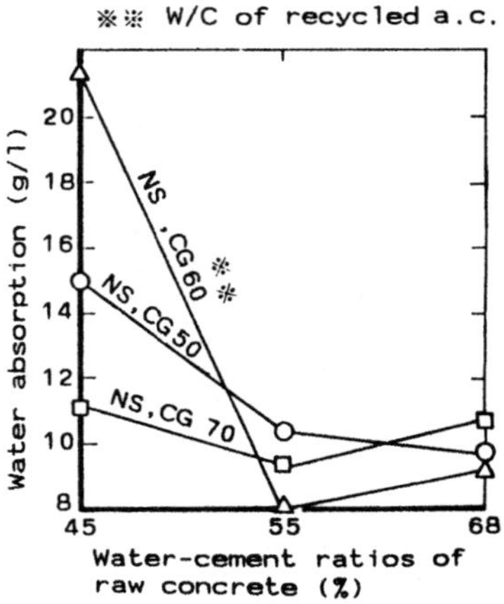

Fig.10 Water-cement ratios of raw concrete of the recycled aggregate concrete vs. water absorption

4.3 Discussions

(a) Observation of water absorption under pressure; Immediately after completion of testing for water absorption under pressure, the test specimens were split and observed. Some data concerning water permeability might be available although the absorbed water is considerably dispersed.

(b) Water absorption of recycled coarse aggregate concrete; Fig 9. shows the relation between the water-cement rations of recycled aggregate concrete and the water absorption. The value of water absorption of recycled aggregate concrete are remained substantially the same even with variable water-cement ratios. The effect of water-cement ratios of original concrete on the recycleded aggregate concrete is shown in Fig.10. The absorption of water became slightly larger for 45 percent while remained substantially the same for 55 and 68 parcent of original concrete. The reason should be as follows the recycled aggeregate made of water-cement ratio of 45 percent contained more amount of mortar and lesser amount of aggregate than that of 55 and 68 percent of original concrete.

5. Conclusions

The following conclusions are derived as a result of this experiments which was carried out with the recycled coarse aggregate concrete replaced by the recycled coarse aggregate 0, 30, 50 and 100 parcent of the crushed stone aggregate.

(1) The compressive strength substantially linearly decreased as the replacement ratio of recycled coarse aggregate increased. The increment ratio with the concrete of a water-cement ratio of 50 percent was smaller than that of the concrete of 70 percent.

(2) The drying shrinkage of recycled coarse aggregate concrete remained substantially the same as the controlled concrete with a replacement ratio lower than 50 percent. When crushed aggregate replaced by recycled aggregate in 100 percent, the shrinkage becomes larger than the controlled concrete.

(3) The depth of accelerated carbonation was remained the same even by changing the replacement ratio of recycled coarse aggegate.

Referring to the BCS results, etc., carbonation rates became sometimes larger when recycled aggregate is replaced by 100 percent.

(4) The resistance to repeated freezing and thawing became smaller as the replacement ratio of recycled coarse aggregate became larger.

It is not recommended to apply recycled coarse aggregate concrete where freezing and thawing damage is severe.

(5) Result of water absorption test under pressure, the absorption of recycled coarse aggregate concrete was 2 - 4 times as large as that of the controlled concrete. With water-cement ratios changed, the rate of water absorption remained substantially the same. In addition, large changes in the water-cement ratio of the original concrete of 55 - 68 parcent did not significantly affect the rate of water absorption of recycled aggregate concrete.

(6) In consequence of above mentioned tests, the recycled coarse aggregate concrete might be used so long as suitable replacement ratio to ordinary aggregate and application site will be selected.

References

The Building Contractors Society of Japan (BCS)(1975-1977), Report of Committee on Disposal and Reuse of Construction Waste (in Japanese).

Kasai,Y.(1975) Water absorption test under pressure for recycled aggregate concrete, Report of Committee on Disposal and Reuse of Construction Wastes, BCS, PP.143-150(in Japanese).

Yanagi,K., Nakagawa,M., Hisaka,M., and Kasai,Y.(1988), Effect of impurities in recycled coarse aggregate upon a few properties of the concrete produced with it, Proceedings of the 2nd RILEM International Symposium on Demolition and Reuse of Concrete and Masonry, Tokyo, Nov. 7-11, 1988.

Yoda,A.(1976) Accelerated carbonation test of recycled aggregate concrete and corrosion test of rebars, Report of Committee on Disposal and Reuse of Construction Wastes, BCS, PP.121-122(in Japanese).

THE DURABILITY OF RECYCLED AGGREGATES AND RECYCLED AGGREGATE CONCRETE

M. MULHERON Department of Civil Engineering
University of Surrey
M. O'MAHONY Department of Engineering Science
University of Oxford

Abstract
The physical and mechanical properties of two recycled aggregates, crushed concrete and clean, graded mixed debris, have been examined and compared with results obtained for a natural river gravel. Initial characterisation tests were performed on both the recycled and natural aggregates to determine their particle size distribution and particle shape, apparent relative density, water absorption and 10% fines value. Tests were then performed to compare the workability, compressive strength, modulus of elasticity, and resistance to freeze/thaw conditions of concretes made with these aggregates. The results indicate that recycled aggregates have lower strengths and densities than natural aggregates. Despite this the durability of concretes manufactured with these recycled aggregates is similar to that of conventional concretes.
Key words: Recycled aggregate concrete, Durability, Youngs Modulus, Compressive strength, 10% fines value, Density, Workability.

1. Introduction

Aggregates for the production of new concrete must meet a number of requirements. Firstly, they must be sufficiently strong for the grade of concrete required and possess good dimensional stability. Secondly, the aggregate must not react with cement or reinforcing steel. Finally the aggregate should have a suitable particle shape and grading to produce a mix with acceptable workability. Based on the results of laboratory investigations and field trials, Hansen (1986), it has been found that clean brick and concrete aggregate can produce a concrete with acceptable workability and strength.

In a previous study by Mulheron (1986) the physical/mechanical properties and durability of dry lean concretes manufactured with recycled aggregates were investigated. The results showed that recycled aggregates were capable of producing lean concretes meeting the compaction and strength requirements of current specifications. In the experiments reported here, the aim was to extend this initial investigation and compare the performance of conventional mass concrete mixes made from natural and recycled aggregates.

2. Experimental methods

Tests have been carried out to assess the mechanical and physical properties of a number of recycled aggregates and concretes made using these aggregates. Additional tests have been carried out on both the loose aggregate and concrete specimens to assess their durability when exposed to freeze/thaw conditions. Tests were also performed on control samples of a natural aggregate and equivalent concrete mixes made with this aggregate.

2.1 Materials under investigation

2.1.1 Aggregates

In this report two types of recycled aggregate have been studied, and their performance compared with that of a natural gravel. The aggregates were; i). Clean graded concrete, ii). Clean graded mixed debris, and iii). Thames valley gravel (and sand) - this irregular flinty gravel was chosen as a 'control' because it had been extensively characterised in previous testing programs.

2.1.2 Concrete

In these experiments the aim was to compare the performance of conventional concrete mixes made from natural aggregates with similar mixes incorporating recycled aggregates as the coarse aggregate. The decision to use only the coarse fraction of the aggregate was based on the finding that the use of recycled aggregate below 2 mm produces inferior quality concrete.

The mix design was formulated to produce a control concrete mix which would allow the water/cement ratio to be varied over the range 0.4 to 0.6 whilst maintaining a workable and cohesive mix. For this reason the aggregate/cement ratio was fixed at 5:1 and the ratio of fine, medium and coarse aggregate was set at 2:1:2. Two mixes were made with each aggregate. The first was designed to be a stiff mix with reasonably low water/cement ratio, w/c=0.45, and having a strength of 55 MPa. The second mix was designed to be more workable, w/c=0.54, with a strength of 40 MPa.

Table 1.1. Mix design, showing quantities of materials used.

Aggregate type	Thames Valley	Crushed concrete	Demolition debris
Coarse (nominal 20 mm)	16.0 Kg	15.70 Kg	14.98 Kg
Medium (nominal 10 mm)	8.0 Kg	7.88 Kg	7.20 Kg
Fine (<5 mm [sand*])	17.0 Kg	17.00 Kg	17.00 Kg
Total aggregate	41.0 Kg	40.58 Kg	39.18 Kg

Cement content = 8.0 Kg (Ordinary Portland cement)
Free water/cement ratio = 0.45 or 0.54
Aggregate/cement ratio = 5:1
* The sand used in ALL the mixes was a Thames Valley gravel

Having graded the aggregates into batches of nominal 20 and 10 mm particle size, it was necessary to adjust the weight of coarse and medium aggregate added to the mix to allow for lower density of the recycled aggregates compared to the Thames Valley control. By manufacturing the recycled aggregates to a specific grading based on equivalent volumes of coarse and medium aggregate a more exact comparison of the control and test mixes could be achieved. The weights of coarse and medium crushed concrete and demolition waste used in the mixes are shown in Table 1.1.

The workability and total water content of each batch of fresh concrete were determined prior to specimen manufacture to ensure that no errors had occurred during the batching process. Cube and prism specimens were then prepared using standard methods and cured at 100% relative humidity and 20°C for a minimum of 24 hours. Further curing of the specimens was carried out in water baths held at 20°C.

2.2 Tests on aggregates

2.2.1 Physical properties

Prior to testing the aggregates were graded by dry sieving following the method described in BS 812 Part 103 (1985). Having obtained the particle size distribution for each aggregate the material was re-combined into the following fractions; Coarse (10-20mm), Medium (5-10mm), and Fine (<5mm).

(a) Relative density

The relative densities of the coarse and medium fractions of each aggregate were determined following the method described in BS 812 Part 2: (1975). In this paper the relative density on an oven-dried basis and the apparent relative density are reported since the ratio of these two values reflects the volume fraction of open voids in the aggregate which can be directly related to the moisture absorption of the aggregate in the saturated surface dry condition.

(b) Moisture absorption

The moisture absorption of an aggregate is defined as the mass of water absorbed into the capillary pores of the saturated surface dry aggregate as a percentage of the dry mass of the aggregate. The weight of water absorbed into the capillary pores of the aggregate can be used to give an indirect measure of the volume fraction of pores present in the aggregate.

(c) 10% fines value

The 10% fines value for the recycled and natural aggregates were determined in accordance with BS 812 Part 3: (1975). This value is the load, in kiloNewton (kN), required to produce 10% fines, defined as material below 2.3 mm, from aggregate particles in the size range 10 to 14 mm. Values in excess of 100 kN are usually required for aggregates for the production of conventional concrete with values in excess of 150 kN being necessary for the production of concrete for hard granolithic floor slabs. Aggregates with a 10% fines value of less than 50 kN would be unacceptable for the production of any cement bound layer or material.

2.2.2 Durability

The durability of the coarse and medium fractions of each aggregate, when subjected to freeze-thaw conditions, was monitored using the following simple test. A known weight of saturated surface-dry aggregate was sealed into a plastic bag with an excess of water and subjected to alternate freezing at -25°C for 24 hours followed by thawing under water at 25°C. After 7 cycles the aggregate was removed from the plastic bag and sieved over a 2.5 mm size sieve and re-weighed. This process was repeated and the weight of aggregate retained on the sieve was monitored as a function of the number of freeze-thaw cycles.

2.3 Tests on concrete

2.3.1 Assessment of fresh concrete

Tests were carried out on each batch of fresh concrete to assess its workability. The properties measured were slump, compaction factor and Vebe time. All tests were performed in accordance with BS 1881 (1983). It should be noted that each of these tests measures a different aspect of the workability of the fresh concrete.

To ensure that the water available to the cement was correct the total water content of each batch of concrete was measured using the 'frying pan' method. By comparing this measured value with the theoretical value from the mix design it was possible to get an indication of the accuracy of the batching and mixing process.

2.3.2 Physical/mechanical properties of hardened concrete

(a) Compressive strength

Testing of the compressive strengths of the concrete cubes was carried out in accordance with BS 1881 Part 116: (1983). The compressive strength was determined on specimens cured for 7 and 28 days. The results reported here are an average of five separate determinations.

(b) Density

The densities of the 28-day old cubes were determined using the method outlined in BS 1881 Part 114: (1983).

(c) Elastic modulus

The dynamic modulus of elasticity, Ed, of the 28 day old concrete prisms was determined by measuring the natural frequency of the specimen when excited using a variable frequency oscillator. The full method being described in BS 1881 Part 5: (1983). In addition the static elastic modulus, Es, was determined following the method described in BS 1881 Part 121: (1983).

2.3.3 Durability of hardened concrete

Testing of the durability of the concrete was carried out using an accelerated freeze-thaw test . Testing was limited to the concretes made with a water/cement ratio of 0.54 since these exhibited the lowest strengths, moduli, and densities they would be most likely to suffer deterioration under freeze-thaw conditions. The test consisted of repeated 48-hour cycles in which saturated surface-dry cubes were sealed into plastic bags containing 1 litre of distilled water or saturated sodium chloride solution and placed into a freezer at -25°C for 24 hours followed by immersion in a water bath at 20°C for a further 24 hours.

To monitor the effect of the freeze-thaw cycles the ultrasonic pulse velocity was recorded at regular intervals using the method described in BS 1881 Part 203: (1986). The direct transmission method was used with the emitter and detector mounted on opposite faces of the cube as this is the most sensitive method for measuring the speed of the pulse travelling through the concrete. To obtain a measure of the overall damage in each cube, the average was taken of the pulse velocity readings taken from all three faces of the cube.

3 Discussion of results

3.1 Aggregates

3.2.1 Physical properties

(a) Relative density

According to Hansen (1986) the density of recycled aggregates can be expected to be somewhat lower than that of equivalent natural aggregates due to the presence of light weight materials such as old mortar and brick. This is confirmed by the results of the tests performed in this study, Table 3.1. It can be seen that the apparent relative densities of the coarse and medium sized recycled crushed concrete aggregates are some 6% lower than those of the Thames valley gravel while those of the recycled demolition debris are 18% lower.

(b) Moisture absorption

Much larger amounts of water are absorbed by the recycled aggregates than the Thames valley gravel, Table 3.1. These results are in close numerical agreement with those of Hansen and Narud (1983) who measured water absorptions for coarse recycled aggregates ranging from 3.7% to 8.7% as compared to the 5.3% to 8.3% obtained here.

Table 3.1 Physical properties of aggregates

	Thames Valley Gravel		Crushed Concrete		Demolition Waste	
	Medium	Coarse	Medium	Coarse	Medium	Coarse
Relative Density (Oven-dried)	2.46	2.52	2.16	2.23	1.89	1.96
Apparent Relative Density	2.55	2.65	2.40	2.48	2.10	2.18
Water Absorption (% of dry mass)	3.1	1.5	8.3	5.4	5.3	5.9
10 % fines value (kN)	180	-	108	-	80	-

Medium = 10 mm; Coarse = 20 mm

(c) 10% fines value

The 10% fines value of the recycled concrete aggregate, Table 3.1, is significantly lower than that of the Thames Valley gravel but still exceeds 100 kN. This indicates that the aggregate has sufficient strength to be viable as an aggregate for the production of mass concrete. The somewhat lower value of 80 kN determined for the recycled mixed debris is less encouraging and would suggest that the use of such aggregates in the production of anything other than low strength concretes would result in the strength of the concrete being limited by the strength of the aggregate.

3.2.3 Durability

The cumulative percentage weight loss of the recycled and natural aggregates are shown in Figure 3.1. The deterioration of the demolition debris aggregate is thought to result from the presence of soft, clay bricks. Such bricks are known to have little resistance to freeze-thaw conditions when tested in the saturated condition. Not all bricks exhibit this behaviour since well burnt, hard bricks are known to be extremely resistant to such damage. The crushed concrete aggregate shows a similar behaviour to that of the Thames Valley gravel but at longer times does show some sign of breaking down.

3.3 Concrete

3.3.1 Assessment of fresh concrete

(a) Workability

The results of the tests to assess the workability of the concrete mixes are shown in Table 3.2. As might be expected increasing the water content in each mix produced an overall increase in workability resulting in an increase in the values of slump and compaction factor and decrease in Vebe time.

The affect of the two recycled aggregates on the workability of the mix can be seen to be quite different. The crushed concrete aggregate produced slightly harsher and less workable mixes than the equivalent control mix as is evidenced by the lower values of compaction factor and slump and longer Vebe times. It is thought that this is a consequence of the particle shape and texture of the crushed concrete aggregate which was both more angular and rougher than the Thames Valley gravel resulting in an increased amount of inter-particle interaction and 'locking'. In contrast the demolition debris derived aggregate produced concrete mixes of similar workability to the control. Interestingly, the individual particles in this aggregate were considerably rounder and less abrasive than the crushed concrete aggregate. This appears to confirm the suggestion that it is the shape and texture of the aggregate particles which is controlling the workability of the fresh concrete.

(b) Total water content

The measured values of total water content in the individual mixes are shown in Table 3.2 along with the theoretical values obtained from the mix design calculations. It can be seen that in all cases there is good agreement between the theoretical and measured values indicating that the batching process employed has been successful in producing concretes with a well controlled free water content.

Table 3.2 Workability and Total water content of fresh concrete mixes

		Thames Valley gravel		Crushed concrete		Demolition waste	
Water/cement ratio		0.45	0.54	0.45	0.54	0.48	0.55
Slump (mm)		20	100	10	20	20	>100
Compaction factor		0.88	0.93	0.84	0.89	0.88	0.96
Vebe time (Seconds)		5.5	3.0	8.0	4.0	6.5	0.5
Total water content (% by weight)	Measured	8.4	9.6	9.9	10.9	10.4	11.3
	Theory	8.2	9.5	10.0	11.2	10.3	11.2

3.3.2 Short term physical and mechanical properties

(a) Compressive strength

The values of the strength achieved by the concrete test cubes are shown in Table 3.3 and it can be seen that the strength of the mixes appears to show some correlation with the type of aggregate in the mix. Thus the mixes with the control aggregate show the highest strengths followed by the crushed concrete and then the demolition debris. It is of note that the strengths of the two concrete mixes incorporating the demolition debris derived aggregate are almost identical despite the difference in free water/cement ratio. This is thought to result from the low strength of the demolition debris aggregate which is acting to limit the maximum strength that the concrete can achieve.

(b) Density

Comparing the densities of the concretes, Table 3.3, it is apparent that the mixes containing the recycled aggregates have lower densities than the control. This is expected since the apparent relative densities of the recycled aggregates are 6-18% lower than that of the control aggregate. The measured density of the concrete made from demolition debris is between 3.5-4.0% lower than the control indicating that the density is lower by an amount proportional to the difference in relative densities of the two aggregates. The mixes containing recycled concrete aggregate show a measured density between 1.5-3.0% smaller than the control concrete which is just within the expected values based on the difference in relative density of the recycled concrete aggregate and the control.

(c) Elastic modulus

From Table 3.3 it is apparent that the recycled aggregate concretes have lower dynamic elastic moduli than the control. The largest difference, 22%, is observed for the concrete containing the recycled concrete aggregate while for the recycled demolition debris the difference is only 16%. These values may be compared with the results of Wesche and Schulz (1982) who found that the modulus of elasticity of recycled aggregate concrete is usually between 15-30% lower than of conventional concrete.

Table 3.3 Physical/Mechanical properties of hardened concrete mixes

		Thames Valley Gravel		Crushed Concrete		Demolition Waste	
Water/cement ratio		0.45	0.54	0.45	0.54	0.48	0.55
Compressive Strength (MPa)	7 days	51.4	40.8	48.8	39.8	33.0	35.6
	28 days	61.5	51.5	59.5	46.4	47.1	46.0
Dynamic elastic modulus - (Ed) (GPa)		47.3	46.7	39.8	36.8	39.0	40.0
Static elastic modulus - (Es) (GPa)		42.4	40.9	30.8	30.0	28.0	32.0
Density at 28 days (kg/m^3)		2372	2383	2336	2312	2274	2300

3.3.3 Durability of the hardened concrete

Figure 3.2 shows the change in measured ultrasonic pulse velocity of the three concretes as a function of the number of freeze-thaw cycles. It can be seen that over the time-scale of the experiment there is a considerable difference in behaviour between the specimens immersed in plain water and those immersed in saturated sodium chloride solution. This is not unexpected since it is known that the presence of de-icing salts leads to an increased rate and extent of attack on concrete exposed to alternate freezing and thawing.

Considering the results for immersion in saturated sodium chloride it is apparent that all of the concretes suffer some damage after 5 cycles and indeed the control mix was observed to fail completely after only 20 cycles. Thus it would appear that the recycled aggregate concretes are showing better resistance to the effects of freeze-thaw conditions than the Thames valley gravel. The results of the specimens immersed in water appear to confirm this finding since whilst the recycled aggregate concrete shows no change in the pulse velocity after 42 cycles the pulse velocity measured in the control mix drops by some 18% over the same number of cycles. This enhanced durability may result from the greater porosity of the recycled aggregates providing a sufficient number of air-filled macro-pores to reduce the pressures resulting from ice formation. It is of note that when subjected to freeze-thaw conditions the recycled aggregates performed less well than the Thames Valley gravel, Figure 3.1. This confirms the findings of Schaler (1930) who found that the results of freeze-thaw test on unbound aggregates do not reliably indicate if a particular aggregate will produce frost resistant concrete.

Given the restricted scope of the test program further testing is required to determine if the recycled and natural aggregate concretes show the same relative performance over a range of conditions.

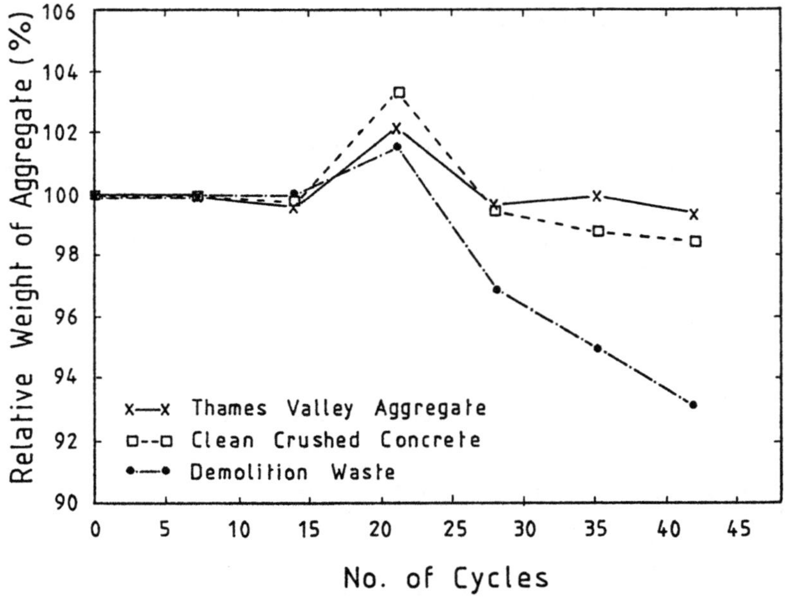

Figure 3.1 Relative weight loss of aggregates as a function of freeze-thaw cycles.

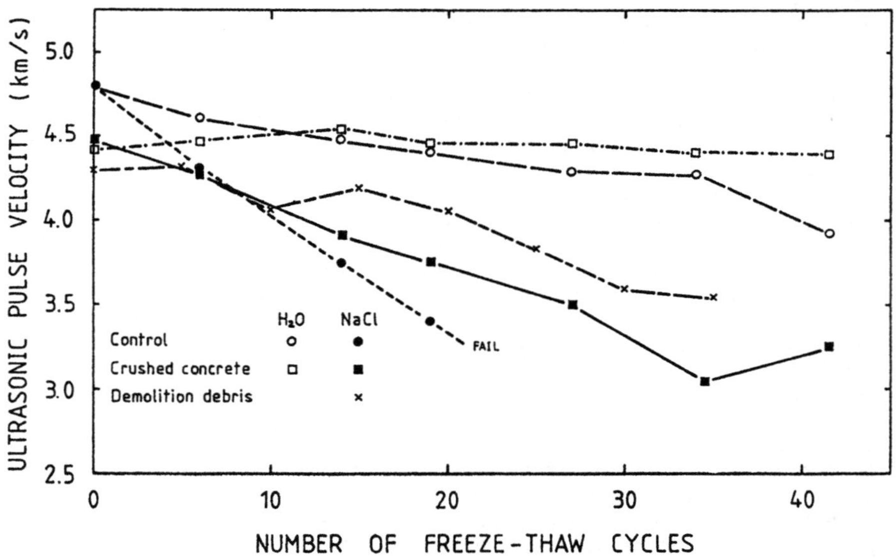

Figure 3.2 Ultrasonic pulse velocity of hardened concrete specimens subjected to alternate freezing and thawing.

4. Conclusions

4.1 Recycled aggregates

1) The physical properties of the recycled concrete and demolition debris aggregates are dissimilar to those of the natural aggregate having properties somewhere between those expected of conventional and lightweight aggregates.

2) The durability of the recycled aggregates exposed to severe freeze-thaw conditions is poor when compared to a natural gravel.

4.2 Recycled aggregate concrete

1) Suitably graded recycled aggregates can be used to produce concretes that satisfy the strength and density requirements of current specifications.

2) The physical and mechanical properties of concrete manufactured using recycled aggregates reflect the porous, low density, low modulus materials present in such aggregates. Thus, when compared to a control concrete, concrete manufactured using recycled aggregates has a lower elastic modulus, lower density, and lower strength.

3) The durability of lean concrete made using recycled aggregates, when subjected to freeze-thaw conditions, appears to be better than, or similar to, an equivalent control concrete made with natural gravel. Further long-term testing is required to confirm this result.

References

HANSEN, T.C., (1986) 'Second state-of-the-art report on recycled aggregates and recycled aggregate concrete', Materials and Structures, September,1986.

HANSEN, T.C. and NARUD, H., (1983) 'Strength of recycled concrete made from crushed concrete coarse aggregate', Concrete International - Design and Construction, 5(1), January 1983.

MULHERON, M., (1986) 'A Preliminary Study of Recycled Aggregates', A report for the Institute of Demolition Engineers, November 1986.

SCHALER, C., (1930) 'Durability of concrete', Proc. Highw. Res. Bd., 10, 132, 1930.

WESCHE, K. and SCHULZ, R. (1982) 'Beton aus aufbereitetem Altbeton - Technologie und Eigenschaften', Beton, 32 (2 and 3).

BS 812 Testing Aggregates. British Standards Institution.
Part 2: (1975) Methods for determination of physical properties.
Part 3: (1975) Mechanical properties.
Part 103: (1985) Method for determining particle size distribution

BS 1881 Methods of testing concrete. British Standards Institution.
Part 5: (1970) Methods of testing hardened concrete for other than strength.
Part 114: (1983) Methods for determination of density of hardened concrete.
Part 116: (1983) Method for determination of compressive strength.
Part 121: (1983) Method for determination of static modulus of elasticity in compression.
Part 203: (1986) Recommendations for measurement of velocity of ultrasonic pulses in concrete.

SOME PROPERTIES OF RECYCLED AGGREGATE CONCRETE

TAKESHI YAMATO, YUKIO EMOTO, MASASHI SOEDA AND YOSHIFUMI SAKAMOTO
FUKUOKA UNIVERSITY, NIPPON STEEL CHEMICAL CO.,LTD.

Abstract
This paper presents the results of an investigation to determine the performance characteristics of concrete produced with recycled aggregates from a plant. The fresh recycled aggregate concrete(RAC) properties studied are slump and air content. The hardened RAC properties investigated are compressive strength, tensile strength, modulus of elasticity and freezing-and-thawing resistance. Test data showed that the recycled aggregate used in this study did not increase the water requirement of concretes. The incorporation of recycled concrete as a new aggregate decreased the compressive strength, the tensile strength and the modulus of elasticity of concrete at 7 to 91 days as compared with those properties of the control concrete. The addition of condensed silica fume, however, improved this reduction of compressive strength due to the recycled aggregate. The freezing-and-thawing resistance of the RAC was lower than that of control concrete of similar composition.
Key words: accelerated tests; air-entrained concretes; compressive strength; freeze-thaw durability; modulus of elasticity; recycled aggregate concretes; condensed silica fume; tensile strength.

1. Introduction

The supply of good quality aggregates has been depleted because of the large construction activity during the past three decades. In addition, there is an aggregate availability problem due to the closing of some aggregate plants and stricter environmental regulations. Disposal of massive quantities of concrete waste raises a difficult problem due to the decreasing availability of dumping areas. Under the above circumstances, the ministry of construction and universities(1976,1983, 1986) in Japan have already started recycling waste materials as construction materials.

Frondistou-Yannas(1980) reported that " The idea of concrete recycling is not new. After the Second World War, Europeans faced with serious waste disposal problems in their destroyed cities turned to demolition debris recycling as a construction material with generally good success ".

The work reported here is the experimental result of an investigation into the performance of the recycled aggregate in the fresh and the hardened concrete.

2. Experimental details

The recycled coarse aggregate used was produced in the first recycling plant in Fukuoka City. The crushed amphibole was used as the natural coarse aggregate. In all the mixes made, ordinary Portland cement was used together with a sea sand. The physical properties of these aggregates are shown in Table 1.

Table 1. Physical properties of aggregates

Properties	Fine aggregate	Coarse aggregate	
		Natural	Recycled
Specific gravity	2.57	2.96	2.47
Absorption(%)	1.32	1.24	4.84
Fineness modulus	2.65	-	-
Maximum size(mm)	-	20	20

The mix proportions used for all the mixes are shown in Table 2. The mixes consisted of non air-entrained concretes and air-entrained concretes with the air-content of 4 and 6 percent. For these concretes, three mixes were made: these included a control mix(N-0) without recycled aggregate and two mixes incorporating the recycled aggregate as 50 percent and full replacement of the natural coarse aggregate. For the freezing-and-thawing test of the air-entrained concrete with the air content of 6 percent, the replacements of coarse aggregate by the recycled materials were 20,30,50,75 and 100 percent. The water to cement ratio of each mix was 0.55 by weight. Both of the natural and the recycled aggregate were pre-soaked for 48 hours in water at 20°C before making them saturated surface dry condition. In all cases the total mixing time was three minutes after the addition of of the mixing water. The slump and the air content measurements were made on all mixes. All test specimens were stored under cover in the laboratory until demoulding at 1 day. After demoulding, they were moved to water at 20°C.

The following properties of the hardened concrete were measured. (1) Compressive strength on 100 mm diameter x 200 mm cylinders, three being tested at each age. (2) Cylinder-splitting tensile strength on 100 mm diameter x 200 mm cylinders, two being tested at each age. (3) Static secant modulus of elasticity on 100 mm diameter x 200 mm cylinders. (4) Freezing-and-thawing resistance on 10x10x40 cm prisms. The specimens for freezing-and-thawing tests were removed from steel forms after one day, and soaked in 20°C water until two to four months after casting. Freezing-and-thawing tests in water were conducted on concrete specimens in accordance with ASTM C 666, Procedure A.

Table 2 Mix proportions and properties of fresh concrete

Type of concrete	Mix no.	W/C (%)	Cement (kg/m^3)	Gr/(Gr+Gn) (%)	Fresh concrete Slump (cm)	Air (%)
Non AE concrete	N-0	55	335	0	9.2	1.6
	N-50	55	335	50	6.7	1.0
	N-100	55	335	100	13.3	1.0
AE concrete	A4-0	55	329	0	9.8	4.4
	A4-50	55	329	50	8.6	4.4
	A4-100	55	329	100	8.9	4.0
AE concrete	A6-0	55	305	0	8.6	7.8
	A6-20	55	305	20	9.0	8.0
	A6-30	55	305	30	15.3	7.0
	A6-50	55	305	50	10.4	6.2
	A6-75	55	305	75	15.5	7.0
	A6-100	55	305	100	14.2	6.2

Gr: Recycled aggregate, Gn: Natural aggregate

3. Results and discussion

3.1 Fresh concrete

The results of the slump values and the air contents measured in the fresh concretes are presented in Table 2. Table 2 shows that the presence of the recycled coarse aggregate leads to a considerable increase in the slump of the mix for a constant mixing water content. This increase of the slump is probably due to the recycled aggregate with a better particle shape than the crushed amphibole. Buck(1973) reports that recycled concrete aggregate does not contain excessive amounts of flat or elongated particles. Malhotra(1978) also reports that the recycled aggregate used in his study was more rounded than crushed limestone and natural sand.

3.2 Hardened concrete

Results for the development of compressive strength up to an age of approximately three months for all mixes are shown in Figure 1. It is evident that the strength continues to increase with age in every case and that the non air-enrained concrete and the air-entrained concrete made from the crushed amphibole have a higher strength than those made from the recycled aggregate at all ages. For the non air-entrained concrete, the compressive strength at 28 days of the RAC with the recycled aggregate of 100 percent is approximately 80 percent of that of the control concrete.

However, the mix of the recycled concrete can be manipulated to obtain the full strength of the control concrete. For instance, one can increase the cement content of the recycled concrete. In this study, the effect of condensed silica fume from a Japanese source on the compressive strength was investigated. The amount of silica(SiO_2) of this condensed silica fume was 89.5 percent by weight. The condensed silica fume was used to replace the cement by 5 to 20 percent by weight. The testing result is shown in Figure 2. Figure 2 shows that the incorporation of condensed silica fume improves the compressive strength of RAC and that the compressive of RAC with 10 or 20 percent silica fume is larger than that of the control concrete. Since the cement plus the silica fume and water contents for these mixes were nearly equal, the improvement in compressive strength for silica fume concrete may be due to the reduction in the water to cementitious

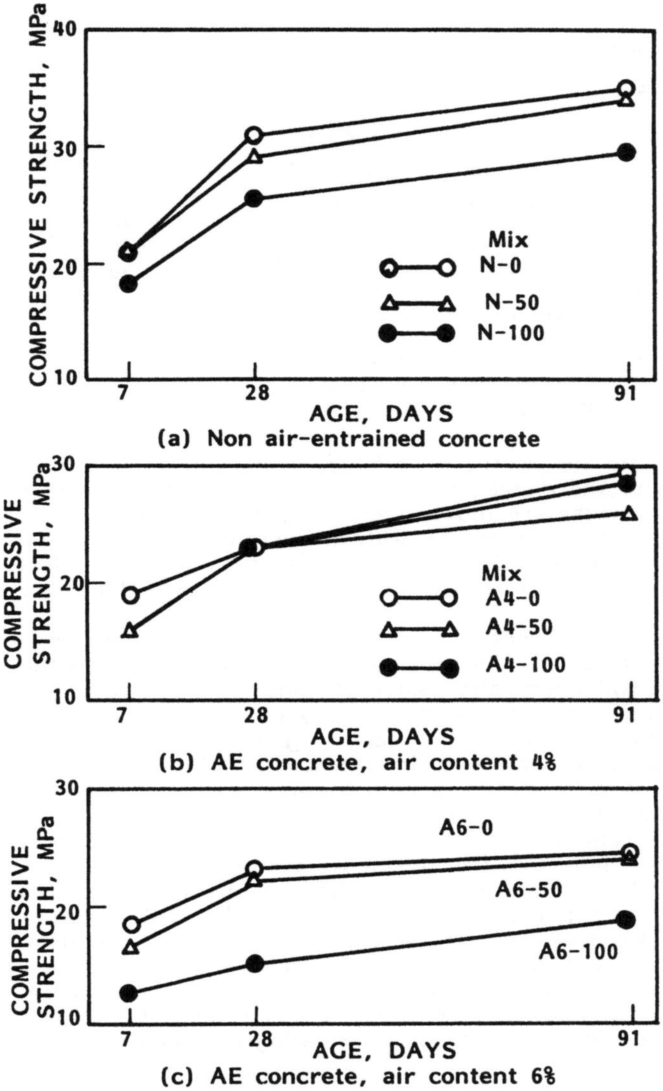

Figure 1. Relationship between compressive strength and age

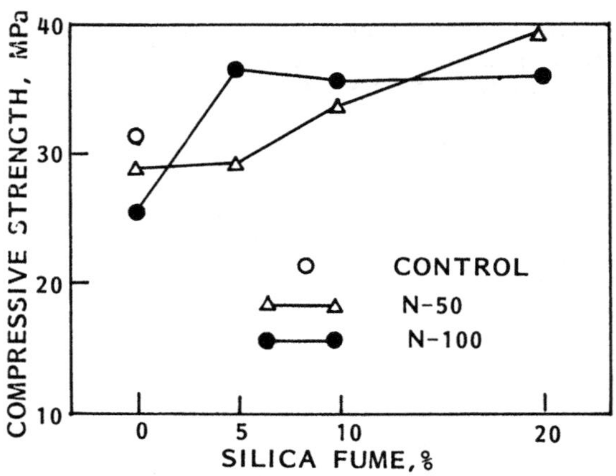

Figure 2. Effect of condensed silica fume on the compressive strength of RAC

Table 3. Durability factors

Mix no.	Durability factor
N-0	8
N-50	5
N-100	4
A4-0	58
A4-50	19
A4-100	4
A6-0	81
A6-20	82
A6-30	65
A6-50	32
A6-75	22
A6-100	13

materials ratio and to the pozzolanic reaction of silica fume.

Results of the development of the indirect tensile strength up to 91 days for all mixes are illustrated in Figure 3. The trends are similar to those for compressive strength discussed earlier. In every case, strength continues to increase with age and Figure 3 indicates that tensile strength is dependent to some extent upon the strength of the aggregate.

The relationship between static modulus of elasticity and age is shown in Figure 4. Figure 4 shows the trend similar to that in corresponding graphs for compressive strength and tensile strength.

The change of modulus of elasticity during freezing-and-thawing cycles is given in Figure 5. Table 3 gives durability factors. Figure 5(a) shows the non air-entrained concretes with and without recycled aggregate have low freezing-and-thawing resistance. For the air-entrained concrete, it is evident from Figure 5(b) and Figure 5(c) that the freezing-and-thawing resistance of the RAC is lower than that of the control concrete. The durability factor of the RAC(A6-100) with the recycled aggregate of 100 percent was 13 in spite of entraining the air content of 6 percent, although that of the control concrete(A6-0) was 81. This reduction in the freezing-and-thawing resistance of the recycled concrete may be due to that the original concrete used as the recycled coarse aggregate probably had low frost resistance. For the replacement by the recycled aggregate less than 30 percent, however, the reduction in the freezing-and-thawing resistance of the RAC was small.

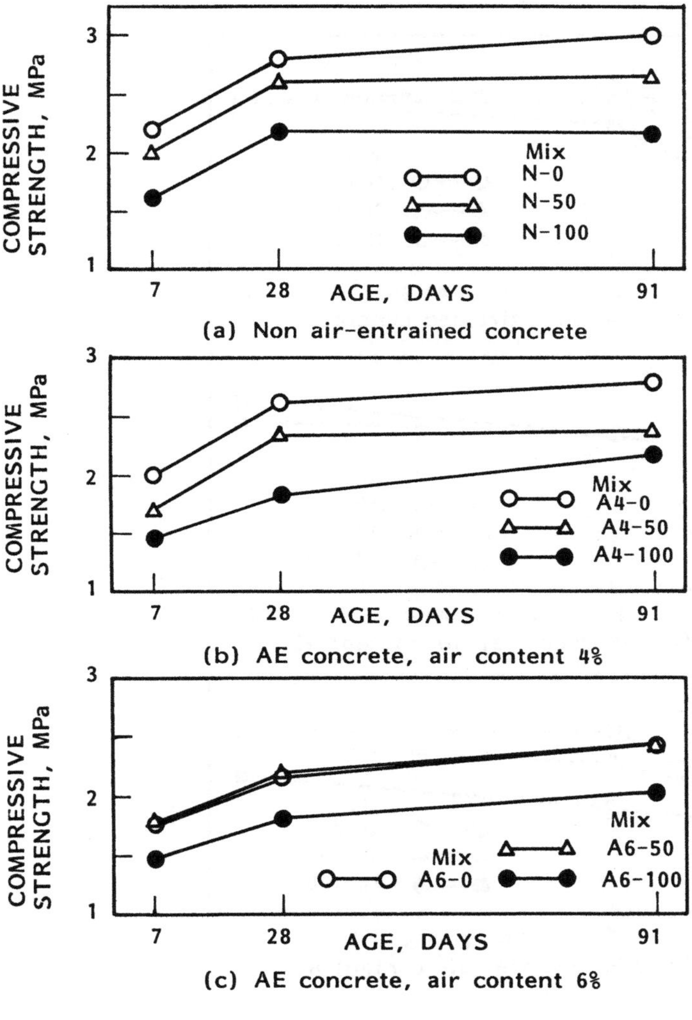

(a) Non air-entrained concrete

(b) AE concrete, air content 4%

(c) AE concrete, air content 6%

Figure 3. Relationship between tensile strength and age

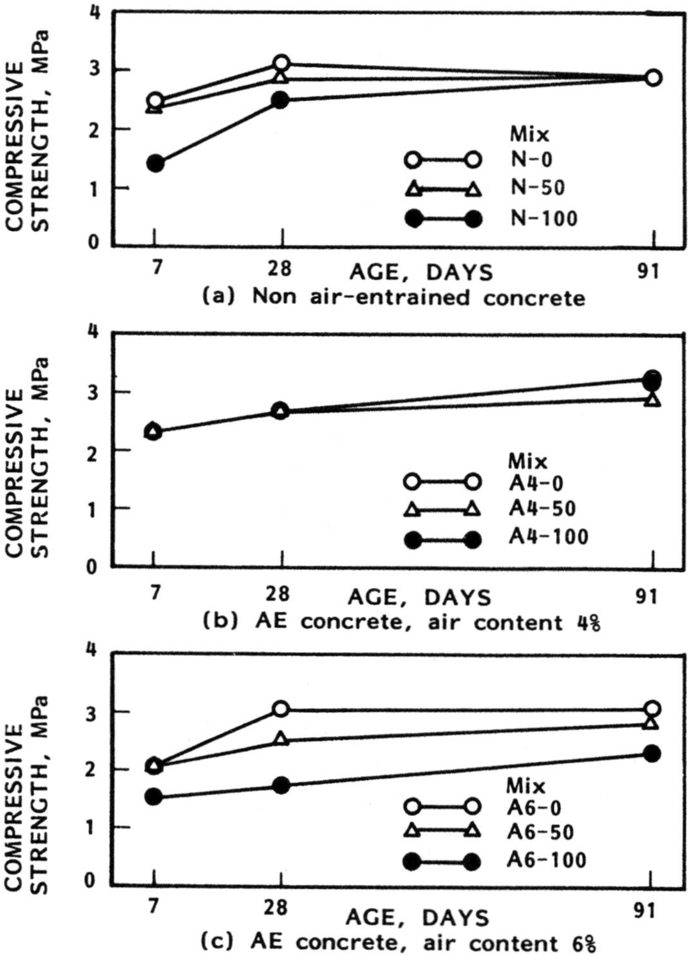

(a) Non air-entrained concrete

(b) AE concrete, air content 4%

(c) AE concrete, air content 6%

Figure 4. Relationship between secant modulus of elasticity and age

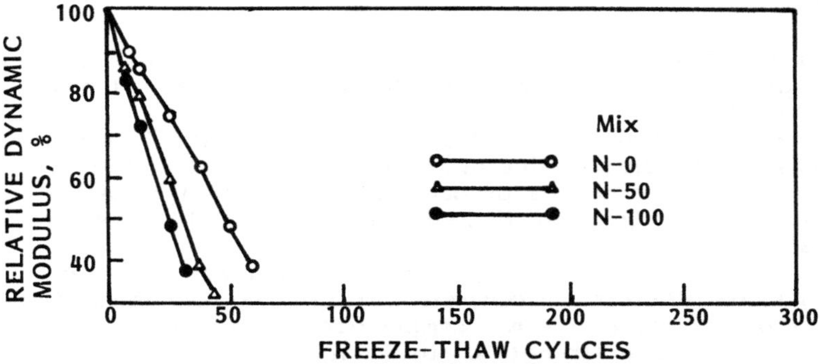

(a) Non air-entrained concrete

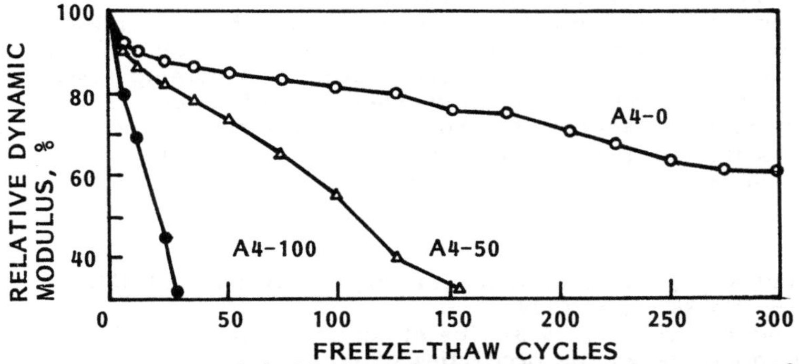

(b) Air-entrained concrete , air content 4%

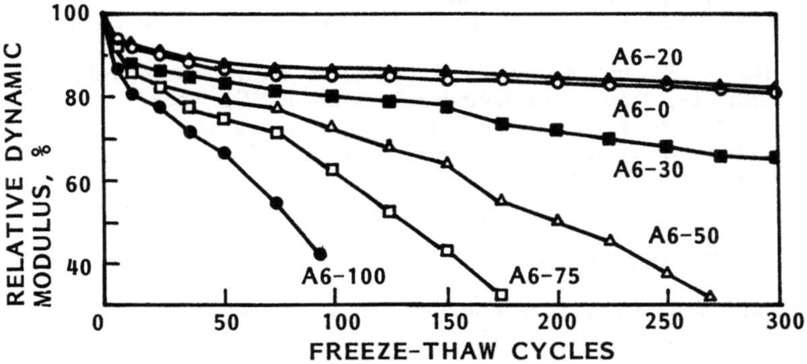

(c) Air-entrained concrete, air content 6%

Figure 5. Number of freeze-thaw cycles versus relative dynamic modulus of elasticity

4. Conclusions

The performance characteristics of concrete produced with the recycled coarse aggregate from a plant are presented in this paper. It has been concluded that:

(1) The presence of the recycled coarse aggregate led to a considerable increase in the slump value of the mix for a constant mixing water content. This increase is probably due to the better particle shape of the recycled aggregate than the amphibole used as the control aggregate.

(2) The recycled aggregate concretes(RAC) had lower strength characteristics than the control concrete with crushed amphibole of the same composition; however, the mix composition can be manipulated to produce RAC of the same strength as the control concrete by the addition of condensed silica fume.

(3) The non air-entrained concretes with and without recycled aggregate had low freezing-and-thawing resistance. The freezing-and-thawing resistance of air-entrained RAC was lower than that of the control concrete. The reduction in the freezing-and-thawing resistance of the RAC was dependent on the proportion of replacement of coarse aggregate by recycled aggregate. For the replacement by the recycled aggregate less than 30 percent, the reduction in the freezing-and-thawing resistance of the RAC was small.

References

Kasai, Y. and Kaga, H.(1976) Reuse of concrete debris, CEMENT & CONCRETE, No.347, pp.20-28.

Mukai, T. and Kikuchi, M.(1983) The effect of replacement of the recycled aggregate on the properties of recycled aggregate concrete (Part 2. The properties of hardened concrete and the effect of steam curing), Proceedings of JASS annual meeting at Hokuriku, pp.85-86.

Shiiba, H., Honda S. and Kitayama H.(1986) A study on recycled aggregate concrete, Report of Kyushu Branch JASS, No.29.

Kawano, H.(1987) Properties of recycled aggregate and recycled aggregate concrete and its usage, CEMENT & CONCRETE, No.490.

Frondistou-Yannas, S.(1980) Recycled concrete as new aggregate, Progress in concrete technology., Canada Centre for Mineral and Energy Technology, pp.639-684.

Buck, A.D.(1973) Recycled concrete, Highway research record, No.430, pp.1-8.

Malhotra, V.M.(1978) Recycled concrete-a new aggregate, Canadian Journal of Civ. Eng., 5, pp.42-52.

MECHANICAL PROPERTIES AND DURABILITY OF CONCRETE FROM RECYCLED COARSE AGGREGATE PREPARED BY CRUSHING CONCRETE

S. NISHIBAYASHI Department of Civil Engineering
Tottori University
K. YAMURA Department of Ocean Civil Engineering
Tottori University

Abstract
The mechanical properties and durability of hardened concrete made from recycled coarse aggregate were studied and compared with those of normal control concrete made from conventional crushed stone.

Recycled coarse aggregate was prepared by crushing several kinds of concrete which were cut out from demolished old concrete structures.

It was found that the compressive strength of recycled concrete was 15 - 30 % lower than that of normal control concrete and that the Young's modulus of the recycled concrete was about 85% of that of the normal control concrete which has almost same compressive strength. These values were mainly affected by the amount of old mortar which remained on the surface of the recycled coarse aggregate.

The resistance to freeze-thaw deterioration of recycled concrete was very much inferior to that of normal concrete, and the resistance was scarcely improved by air entrainment. Durability of the recycled concrete under sulphate action and sea water was almost the same as that of normal concrete. The durability properties of recycled concrete were considerably affected by the water-cement ratio.
Key words: Recycled aggregate concrete, Compressive strength, Creep, Drying shrinkage, Durability

1. Introduction

The study of recycled aggregate concrete in which demolition waste is utilized to produce aggregate for new concrete, can contribute to the solution of two problems. The first is the shortage of aggregate from river, and the second is the waste disposal problem.

Recent investigations concerning recycled aggregate concrete have shown that concrete made from crushed concrete has qualities not much inferior to those of concrete from normal crushed stone, and that it has the possibility to be in common use except in cases where special high quality concrete is required. However, there is still many unknowns concerning the basic properties of recycled aggregate concrete, and this prevents the wide practical use of this kind of concrete. In this project, the authors have studied the elastic-plastic properties and the durability of recycled concrete which has been made fromthe coarse aggregate of crushed old concrete.

2. Experiment outline, preparation and properties of materials

2.1 Materials

Normal portland cement and river sand, of which the specific gravity and the fineness modulus were 2.60 and 2.80, respectively, were used.

Recycled coarse aggregate was prepared by crushing several kinds of old concrete. Some of the old concrete was manufactured in the laboratory for this study, and some was cut out from old concrete structures, such as the concrete piers of old bridges, concrete dams and reinforced concrete buildings. Properties and histories of the old concrete are shown in Table 1. Recycled aggregate was produced by a jaw crusher which was set to produce coarse aggregate of 25mm, at maximum. The crushed products were classfied into three size groups: In excess 30 mm, 5-30 mm and below 5 mm. The group of 5-30 mm was used as the recycled aggregate. Particles greater than 30 mm were recrushed. The fine part, below 5 mm, was waste. The physical properties of the recycled aggregate are shown in Table 1, together

Table 1 Properties and histories of the original concrete and physical properties of the aggregate

Symbol		R-1	R-2	R-3	R-4	R-5	R-6	R-7	C
Date of manufacture		1979			1959	1966	1935	1932	-
Strength (MPa)		23.4	40.6	49.6	17.6	28.4	11.8	22.5	
Coarse aggregate *		C.S.			R.G.	C.S.	C.S.& R.G.	R.G.	
Used purpose		Manufactured for this study			Low RC building		Dam	Bridge (RC)	
Max. size (mm)		25	25	25	25	25	25	25	25
Specific gravity		2.43	2.43	2.43	2.45	2.45	2.36	2.32	2.70
Water absorption(%)		7.0	6.9	6.8	5.7	7.2	8.1	7.3	0.8
Fineness modulus		6.95	6.96	7.02	7.36	7.25	7.18	7.11	7.11
Bulk density(kg/m^3)		1310	1310	1310	1420	1370	1360	1340	1540
Percent of solid volume(%)		54	54	54	58	56	58	58	-
Crushing strength	Crushed value 40t(%)	24.6	23.1	23.0	30.6	33.4	-	27.4	-
	Fineness value (10%)	11.3	13.0	13.3	7.3	6.0	-	9.3	-
Mortar content %		35.5	36.7	38.4	40.0	50.0	67.6	43.4	-

* C.S. : Crushed stone, R.G. : River gravel

with those of crushed stone which was used for the production of normal control concrete(C).

2.2 Testing plan and conditions

The scope of the tests performed is summarized in Table 2.

The mix proportions of the normal and recycled concrete, with a constant slump of 7.5 cm, were determined in trial batches.

2.3 Experiment procedures

(1) Mechnical properties

The compressive strength and Young's modulus were measured for the cylindrical specimens of ø 10 x 20 cm, which had been cured in water at 20 °C for 28 days.

The concrete specimens used for the creep test were two prisms of 10 x 10 x 38 cm which were arranged in series, with metal plates set on both sides and between the prisms. A load for introducing sustained stresses was applied to the specimens with a prestressing rod (ø 13 mm) placeing at the center of the section of each specimen. Prior to the loading, the specimens were cured in water at 20 °C for 28 days. After loading, the specimens were placed in air of 20 °C with the R.H. at 70%. Loading stress levels for the creep test were 0 (for drying shrinkage), $1/4 f'_c$ and $1/3 f'_c$, where f'_c was the compressive strength of concrete at the age of loading. Recycled aggregate used in these tests was R-2.

(2) Durability tests

Freeze-thaw durability tests were carried out according to ASTM C666A (Resistance of Concrete to Rapid Freezing and Thawing, Standard Method of test for,).

Table 2 Testing plan

(a) Test for mechnical properties

Testing elements	Testing conditions
Aggregate	Crushed stone : C Recycled aggregate : R-1 - R-9
Unit cement content (kg/m³)	250, 350, 450
Water cement ratio (%)	43 - 62
Tests	Compressive strength Young's modulus Drying shrinkage & creep

(b) Tests for durability

Testing elements	Testing conditions
Aggregate	Crushed stone : C Recycled aggregate : R-7
Water cement ratio (%)	45, 65
Admixture	Not including (PL) Water reducing agent (AE type) (AE)
Tests	Freeze-thaw test Sulphate durability test

To evaluate the durability of concrete against sulphates, the accelerated test described below was carried out. Concrete prisms of 10 x 10 x 40 cm were subjected to repeated cycles of immersion in a testing bath (a mixed solution of 20% Na_2SO_4 and $MgSO_4$ at 20°C) and drying in an oven(70°C). The period of immersion and drying were 24 hours each per cycle. This test was started at the age of 14 days. Until the begining of the test, the specimens were cured in water at 20°C. To evaluate the degree of deterioration, the reduction in the dynamic modulus of elasticity and the change in weight and length were measured at the end of each cycle.

3. Results and discussions

3.1 Compressive strength and Young's modulus

Relationships between the compressive strength and the cement water ratio of the recycled concrete are shown in Fig. 1, compared with those of normal control concrete. The compressive strength of recycled concrete is 15 - 30 % lower than that of normal control concrete. Especially at the high strength range, such as above 30 MPa, the compressive strength of recycled concrete hardly increases with a decrease of the water cement ratio. Fig 2 shows the relationship between the coefficient of variation in the compressive strength and the cement water ratio. Each value of the coefficient of variation was obtained from the measured values of 3 - 5 specimens. For the normal control concrete, the accuracy of the tests was estimated to be moderately good, however the coefficient of variation for the recycled concrete was quite large. This shows that the compressive strength of recycled concrete varies widely, and that substantial consideration is necessary to decide the specified design strength.

Fig. 3 and 4 show the relationship of the compressive strength of recycled concrete to the mortar content of recycled aggregate, and to

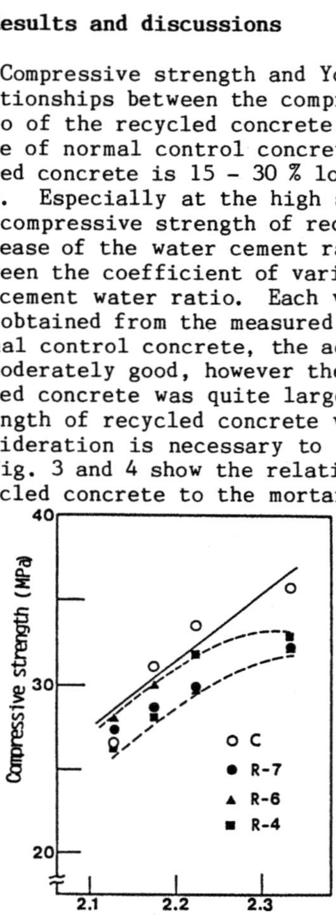

Fig.1 **Relationship between the compressive strength and the cement water ratio of concrete**

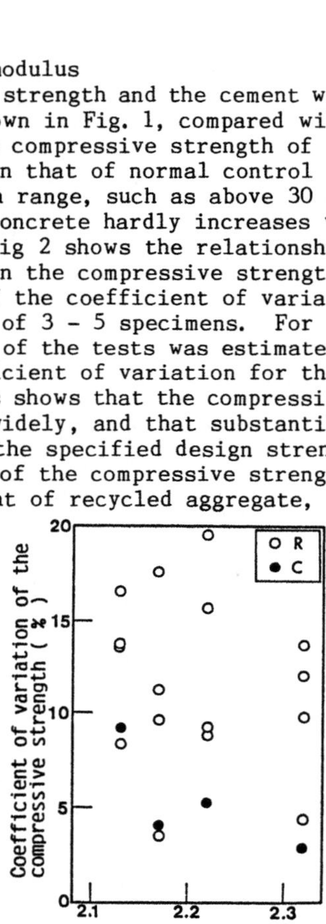

Fig.2 **Coefficient of variation of the compressive strength of concrete**

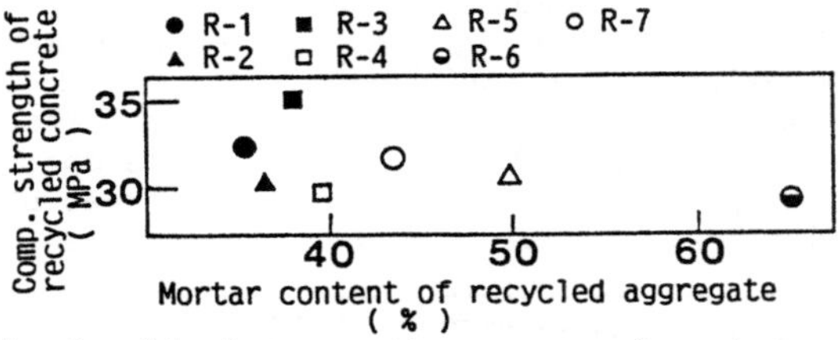

Fig. 3 Relationship between the compressive strength of recycled concrete and the mortar content of the recycled aggregates

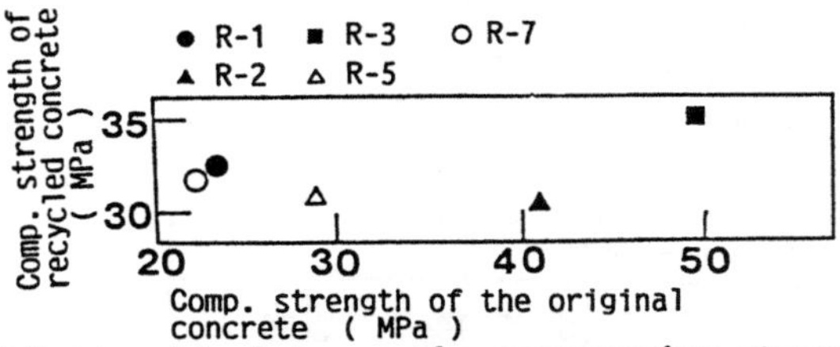

Fig. 4 Relationship between the compressive strength of the recycled concrete and that of the original concrete

the compressive strength of the original concrete, respectively. From These figures, the compressive strength of recycled concrete is scarcely affected by parameters such as mortar content or compressive strength of original concrete, which are considered to indicate the properties of recycled aggregate.

The relationship between the compressive strength and Young's modulus is shown in Fig. 5. For both the recycled concrete and the normal concrete, linear relationships between the compressive strength and the Young'smodulus are observed. The Young's modulus of the recycled concrete is 65 - 85% of that of normal control concrete which has almost same compressive strength. These values are affected by the properties of the recycled aggregate, such as the old mortar content.

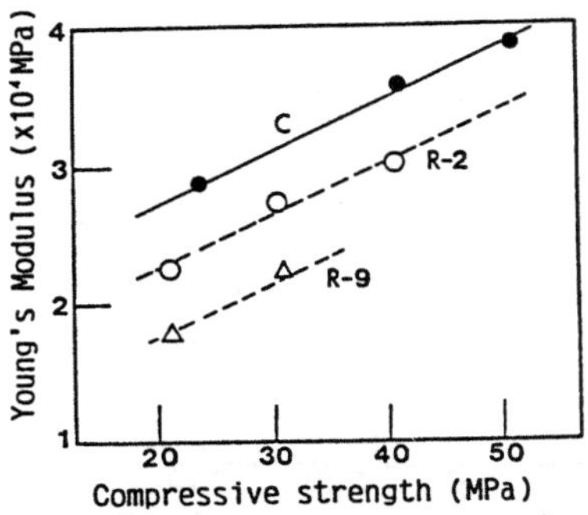

Fig. 5 Relationship between the Young's modulus and the compressive strength of concrete

3.2 Drying shrinkage and creep characteristics

Time - drying shrinkage curves for the recycled concrete are shown in Fig. 6, with the measuring of the shrinkage starting at 28 days of concrete age. The specific creep strain - time curves after loading (at 28 days of concrete age) are shown in Fig. 7. The specific creep strain is defined as $\varepsilon p/\sigma$, where εp is the total creep strain and σ is the sustained loading stress in a concrete specimen.

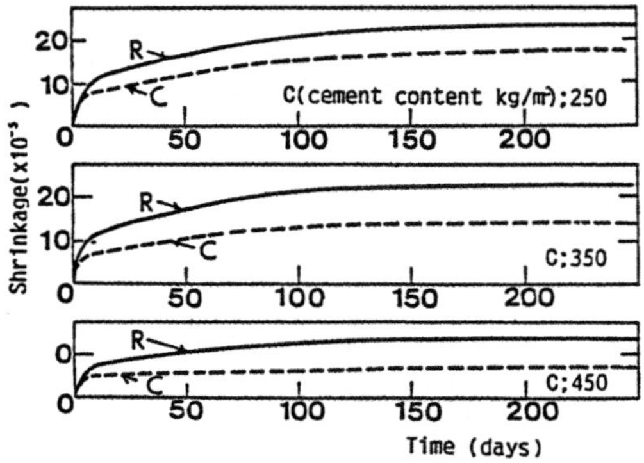

Fig. 6 Relationships between the drying shrinkage and time

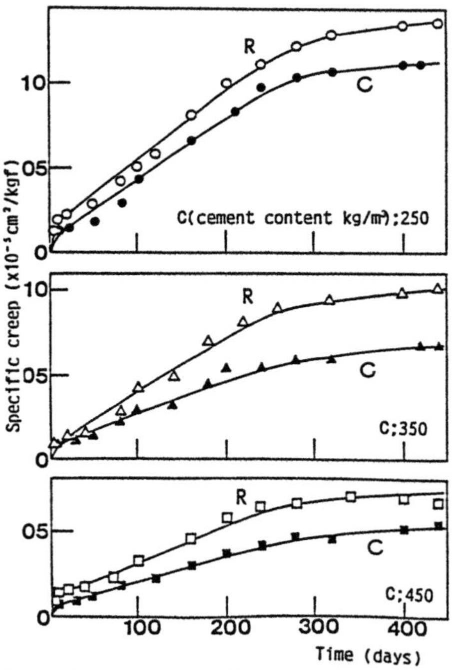

Fig. 7 Relationships between the specific creep and time(1/3 f'_c)

The drying shrinkage of recycled concrete is considerably larger than that of normal control concrete, except at the very early stage. This is attributed to the recycled aggregate containing large amounts of old mortar, with many pores. This induces a greater shrinkage, and makes the total shrinkage of the recycled concrete greater than that of normal control concrete.

The specific creep of recycled concrete is greater than that of normal concrete. The differences developed over a period of 250 - 300 days after loading. Then, the rate of increase in creep strain for both concretes became gradually smaller with a further increase in time. Creep strain increased considerably with an increase of the water cement ratio. The difference in creep strain between these two kinds of concrete is almost constant in any water cement ratio. Since the specific creep strain of recycled concrete is also constant regardless of the sustained loading level, the Davis-Glanville law can be applied to recycled concrete. It is considered that these creep characteristics are largely affected by the old mortar contained in the recycled aggregate.

3.3 Durability of recycled concrete

The results of the freeze-thaw durability tests are shown in Fig. 8. E/E_0 in this figure indicates the ratio of the dynamic modulus of elasticity at the end of each cycle to the initial value. Similarily, W/W_0 represents the ratio of the weight change in the specimens. It is clear from Fig.8 that the freeze-thaw durability of recycled concrete is very much inferior to that of normal concrete, and that this durability can be scarcely improved by air entrainment. It is considered that the old mortar contained in the recycled aggregate is easily damaged by repeated freeze-thaw action. The freeze thaw durability is affected by the water cement ratio of the concrete, too. Therefore, the combined behaviour of the old and new mortar is considered to have a notable effect on the freeze-thaw durability of recycled concrete.

Fig. 9 shows the relationship between the number of cycles of immersion in sulphate solution and oven-drying to the changes in the dynamic modulus of elasticity, the weight loss and the length changes of the specimens. From Fig. 9, it was found that the durability under sulphate action of recycled concrete is equal or slightly inferior to that of normal concrete

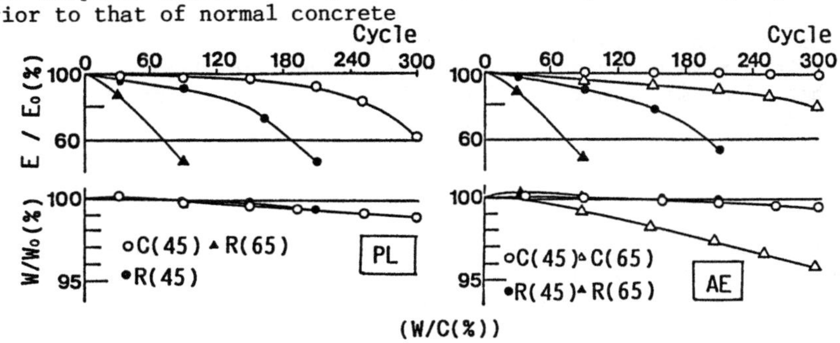

Fig. 8 Freeze-thaw durability of concrete

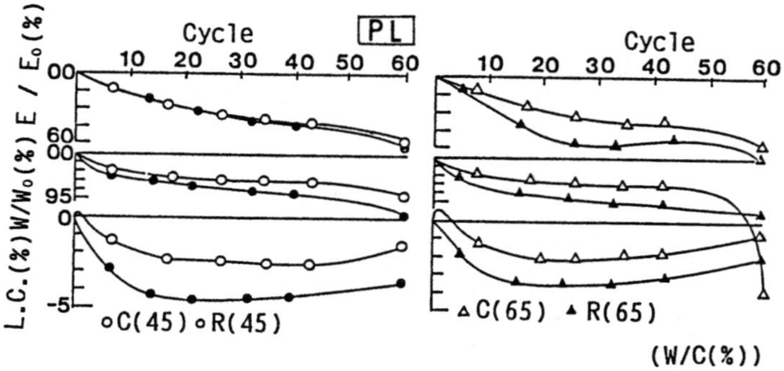

Fig. 9 Durability under action of sulphate

4. Conclutions

The purpose of this study is to clarify the mechnical properties and durability of recycled concrete using crushed waste concrete in place of conventional coarse aggregate. The conclusions of the present study may be summarized as follows:-

(1) The compressive strength of recycled concrete is 15 - 30% lower than that of normal concrete which has the same water cement ratio. In the high strength range, such as above 30 MPa, this compressive strength hardly increases with a decrease in the water cement ratio.

(2) The compressive strength of recycled concrete varies widely. In this experiment, there were cases where this value of recycled concrete was about 20 %.

(3) Young's modulus of recycled concrete is 65 - 85% of that of normal concrete which has the same compressive strength. A linear relationship was found between the compressive strength and Young's modulus.

(4) The drying shrinkage and the specific creep strain of recycled concrete, at the final stage, becomes considerably larger than that of normal concrete. The Davis-Granville law can be applied to recycled concrete under a low sustained stress level such as below 1/3 of compressive strength of the concrete.

(5) The freeze-thaw durability of recycled concrete is very inferior, and can not be improved by air entrainment. The durability of recycled concrete under sulphate action is equal or slightly inferior to that of normal concrete. These durability properties of recycled concrete are considerably affected by the water cement ratio of the concrete.

References

1) The Commitee for the Disposal and Recycling of Construction Waste (The Society of Building Constructors), "Study on Recycled Aggregate and Recycled Concrete," Concrete Journal, Vol. 16, No. 7, July 1978.

2) Nishibayashi, S. and K. Sakata, "A study on the Accelerated Test for Durability of Concrete under Action of Sulphate," Proc. of JSCE, No. 207, Nov., 1972.

STRENGTH AND DRYING SHRINKAGE BEHAVIOR OF CONCRETE USING CONCRETE CRUSHED AGGREGATE

TADAYOSHI FUJII

Institute of Technology
Shimizu Corporation

Abstract
This paper describes the possibility to reuse concrete waste produced by demolition of reinforced concrete structures as aggregate for concrete from the viewpoint of strength, drying shrinkage and weight loss. Concrete rubble obtained from the demolished buildings were crushed by various crushing machines to reuse as aggregate.

The strength, drying shrinkage and weight loss of recycled concrete were examined experimentally when the types and combinations of aggregate, water cement ratio, slump, curing method and so on were varied.

Drying shrinkage of recycled concrete which used concrete crushed aggregate showed larger value by 20-30% than conventional river sand-gravel concrete. But their general behaviors are similar to the conventional concrete, and recycled concrete seems to have no problem for their practical use. But as CS·CG concrete shows less strength and larger shrinkage and weight loss, water permeation is considered to be large.

Therefore in using CS·CG concrete practically, care should be taken.
Key Words: Original concrete, Recycled concrete, Concrete crushed aggregate, Concrete crushed sand, Concrete crushed gravel.

1. Introduction

In Japan, it is proposed that the demolished concrete should be crushed and reused as aggregate of concrete from a viewpoint of scarcety of natural aggregate and conservation of energy at producing aggregate.

From this idea, in 1974, "Committee on Treatment and Reuse of Concrete Waste" was established in BCS (Building Contractors Society) and started their activities until 1977. The Committee carried out a wide range of studies and this paper reports a part of experimental rusults of the joint study of committee members.

The test were carried out by Shimizu Corporation, Meiji University and Nihon University as shown in Series I, II and III respectively.

2. Experimental Program

2.1 Purpose

To measure the strength, drying shrinkage and weight loss of recycled concrete which used concrete crushed aggregate, test was carried out by the following combination of factors as shown in Table 1.

Table 1. Combination of various factors

Item	Series I	Series II	Series III
Type of aggregate	Natural aggregate 1 (NA) Crushed aggregate 5 (CA)	Natural aggregate 1 Crushed aggregate 4	Natural aggregate 1 Crushed aggregate 4
Combination of aggregate	NS·NG S:sand NS·CG G:gravel CS·CG	NS·NG NS·CG CS·CG	NS·NG NS·CG (NS+CS)·CG
W/C(%)	45 50 60 70	50	50 60
Slump (cm)	21	5 20	8 21
Curing method	Water curing Curring room exposure	Curing room exposure water curing 1 curing room exposure 3 Steam Curing curing room exposure 1	Indoor exposure
Measuring method of shrinkage	Dial gauge (free shrinkage)	Comparator (free shrinkage restrained shrinkage)	Comparator (free shrinkage)

2.2 Material

The properties of cement and aggregate in Series I are shown in Table 2 and 3. The crushed aggregate was made by crushing 23 years old concrete rubble obtained from a demolished reinforced concrete building. Various crushing machines such as Jaw crusher, Threader, Cone crusher and so on were used. Coarse and fine aggregate were classified by screens.

The strength of original concrete was supposed to be 250-300 kgf/cm^2 by chemical analysis of fragments.

As admixtures, air entraining agent was used.

Table 2. Physical properties of cement (Series I)

Density	Specific surface area (cm^2/g)	Setting time (hour-minute)		Flow (mm)	Bending strength (kgf/cm^2)		Compressive strength (kgf/cm^2)	
		Initial	Final		7day	28day	7day	28day
31.5	3160	2-29	3-43	242	46.0	71.4	230	411

Table 3. Symbol, type and combination of aggregate (Series I)

Aggregate type	Symbol	Source or type of crusher	Density saturated surface-dry condition	Absorption ratio (%)	Unit weight (kg/l)	Unit weight / Density (%)	Maximum size (mm)	Fineness modulus
Fine	NS	Fuji river	2.58	2.1	1.69	65.5	2.5	2.61
	CS(M)	Cone mill	2.31	8.7	1.48	64.1	5.0	2.60
	CS(G)	Jaw&threader	2.25	8.8	1.42	62.0	5.0	3.51
	CS(I)	Threader	2.25	10.6	1.40	62.2	5.0	2.86
	CS(N)	Jaw	2.22	11.4	1.44	64.9	5.0	3.18
Coarse	NG	Naka river	2.60	1.9	1.66	63.8	25.0	6.79
	CG(M)	Cone mill	2.45	4.6	1.37	55.9	25.0	6.79
	CG(G)	Jaw&threader	2.39	5.8	1.38	57.7	50.0	7.24
	CG(H)	Jaw&threader	2.38	5.3	1.38	58.0	50.0	7.11
	CS(I)	Threader	2.44	4.9	1.42	58.2	20.0	6.51
	CG(N)	Jaw	2.42	5.7	1.29	53.3	15.0	6.52

Note: Symbol in ( ) shows the different crushing method

2.3 Mix proportion

Mix proportion of concrete is shown in Table 4. In series I, same mix proportion for the same W/C and S/A was used, although the combination of aggregate varied. As a result, slump varied from 12cm to 21cm and concrete with large quantities of crushed aggregate showed lower slump. In Series II and III, to obtain the intended slump value water content and S/A were varied.

Table 4. Mix proportion of concrete

Series	Combination of aggregate	Slump (cm)	Air (%)	W/C (%)	S/A (%)	Unit water (kg/m³)	Volume (t/m³) cement	Volume (t/m³) Fine aggregate	Volume (t/m³) Coarse aggregate	Weight (kg/m³)	Weight (kg/m³) Fine aggregate	Weight (kg/m³) Coarse aggregate
I	NS·NG	12		45		215	152	237	356	478	604	903
	NS·CG		4	50	40	209	133	247	371	418	630	946
	CS·CG	21		60		204	108	267	401	340	681	1023
				70		204	92	274	410	291	699	1046
II	NS·NG				46.7	156	99	323	368	312	846	975
	NS·CG	5	5	50	46.8	162	103	320	364	324	838	895
	CS·CG				46.9	162	103	321	363	324	732	893
	NS·NG				48.8	205	130	300	315	410	786	835
	NS·CG	20	3.5	50	46.5–49.1	197–209	125–133	294–304	312–338	394–418	770–796	755–828
	CS·CG				49.1–49.3	194–203	123–129	286–299	297–309	388–406	652–688	731–751
III	NS·NG				42.1	147	93	303	417	298	806	1105
	NS·CG	8	4	50	40.0–45.8	162–168	103–107	277–318	378–408	324–336	737–846	926–991
	(NS+CS)·CG				45.8	162	103	318	378	324	821	937
	NS·NG				48.7	186	98	329	347	307	875	920
	NS·CG	21	4	60	44.1,51.7	199,204	105,108	302,339	317,346	331,346	803,902	777,841
	(NS+CS)·CG				51.7	199	105	339	317	331	866	777

2.4 Experimental method

1) Mixing, moulding and curing

After mixing, concrete is poured into steel moulds of 10x10x40cm and specimens were demoulded next day. After demoulding, concrete specimens were cured in various ways; they were

(1). Curing room (20°C, 50%) exposure. (Series I,III)

(2). Water cured for 1 week and curing room exposure afterward. (Series I,II)

(3). Curing room exposure with relative humidity (RH) of 50% for 5 weeks and 60% afterward. (Series II)

(4). Indoor exposure under varying temperature (10-31°C) and relative humidity (53-84%). (Series II)

As for the compressive and tensile strength test, concrete was poured into cylinder moulds of 10cm diameter by 20cm height.

After demoulding, specimens were cured in 20°C water or curing room. (Series II)

2) Measuring method

Compressive and tensile strength of recycled concrete was tested according to JIS (Japanese Industrial Standard) A 1108; "Testing method of compressive strength of concrete" and JIS A 1113; "Method of testing tensile strength of concrete" (splitting test) respectively.

At compressive strength test, Young's modulus was also measured by Compressometer.

Measuring of drying shrinkage was carried out by JIS A 1129; "Testing method of length change of mortar and concrete".

Weight of specimen was measured at the same time.

3. Results and discussion

3.1 Compressive strength, tensile strength and Young's modulus

The compressive strength, tensile strength and Young's modulus of recycled concrete is shown in Table 5. NS·NG concrete showed the largest compressive, tensile strength and Young's modulus, while CS·CG concrete showed the lowest strength behavior by about 20-30% lower than NS·NG concrete.

The density of recycled concrete was smaller than NS·NG concrete, and the density of NS·CG, CS·CG concrete was about 95%, 90% of that of NS·NG concrete respectively. This is due to the smaller density of concrete crushed coarse and fine aggregate used.

3.2 Drying shrinkage of recycled aggregate

1) Influence of W/C and slump on shrinkage (See Fig. 1,4,5)

Drying shrinkage of concrete was $7\text{-}12\times10^{-4}$ at the age of 13 week and $8\text{-}12\times10^{-4}$ at 31 week and they were still in progress.

As the W/C of concrete becomes larger, the drying shrinkage also becomes larger. (See Fig. 1)

When the shrinkage of concrete of slump 8cm is compared with that of slump 20cm concrete, concrete with larger slump showed a bit higher shrinkage. (See Fig. 4)

But shrinkage of slump 5cm concrete showed smaller shrinkage by about 10% than slump 20cm concrete. (See Fig. 5)

2) Influence of aggregate type and combination (See Fig. 1,3,4,5)
The shrinkage difference by different aggregate type caused by different crushing methods was $2\text{-}3 \times 10^{-4}$ at 13 week.

Concrete which used Type G aggregate that was crushed twice by jaw crusher first and secondary crushing by threader showed smaller shrinkage than those of other concretes which used aggregate crushed one time. (Type H,I,M,N) This decrease of shrinkage is considered to be due to the more removal of adhered mortar on aggregate surface by crushing.

The order of aggregate combination which shows the smaller shrinkage to larger one is as follows: NS·NG < NS·CG < CS·CG.

NS·NG concrete showed the smallest shrinkage, while CS·CG concrete showed the largest.

The more crushed aggregate used, the bigger the shrinkage becomes.

Table 5. Strength behavior and density of recycled concrete (age:28day)

Item	Compressive strength (kgf/cm^2)			Tensile strength (kgf/cm^2)			Young's modulus $E_{1/3}$ ($\times 10^5 kgf/cm^2$)		Density (kg/l)
Slump	20		5	20		5	20		20
Curing / Aggregate combination	NO Water (20°C)	N1 Room (20°C, 50,60%)	NO Water (20°C)	NO Water (20°C)	N1 Room (20°C, 50,60%)	NO Water (20°C)	NO Water (20°C)	N1 Room (20°C, 50,60%)	N1 Room (20°C, 50,60%)
NS·NG	----	266 (100)	329	----	29.8 (100)	37.8	----	2.55 (100)	2.24 (100)
NS·CG(M)	289	258 (97)	----	34.7	26.0 (87)	----	2.00	1.95 (76)	2.13 (95)
NS·CG(G)	----	274 (103)	----	----	27.7 (93)	----	1.98	1.95 (76)	2.14 (96)
NS·CG(N)	----	248 (93)	324	----	27.4 (92)	36.1	----	1.73 (68)	2.12 (95)
NS·NG(I)	284	252 (95)	----	32.7	28.3 (95)	----	2.06	1.98 (78)	2.14 (96)
CS(M)·CG(M)	220	227 (85)	----	28.4	24.4 (82)	----	1.65	1.69 (66)	2.00 (89)
CS(G)·CG(G)	----	189 (71)	----	----	21.9 (73)	----	1.68	1.60 (63)	2.02 (90)
CS(N)·CG(N)	----	176 (66)	309	----	22.1 (74)	34.3	----	1.38 (54)	1.95 (87)
CS(I)·CG(I)	198	186 (70)	----	26.0	18.9 (63)	----	2.17	1.78 (70)	1.95 (87)

Note : (1) Value is an average of three test specimens, and value in ( ) shows the ratio to NS·NG concrete.
(2) Curing method, NO: water cured (20°C), N1: room cured 0-5W:20°C,50%RH
5W- :20°C,60%RH

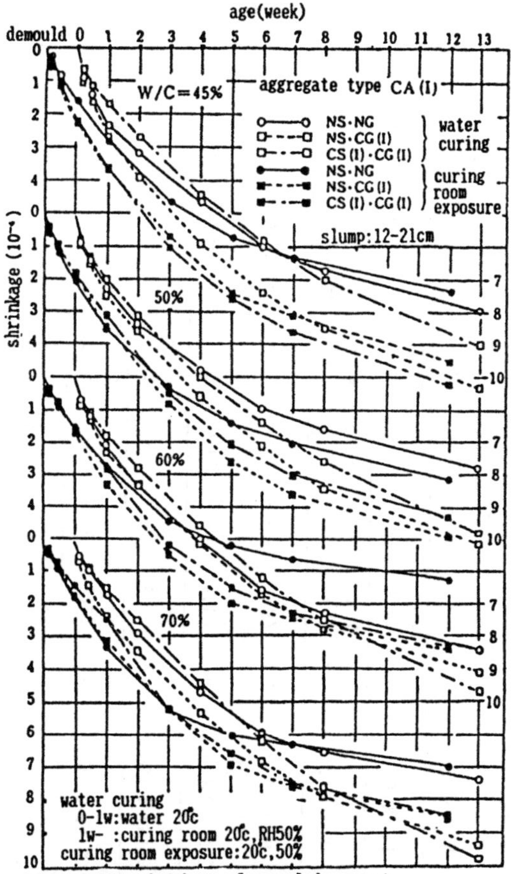

Fig.1 Drying shrinkage of recycled concrete
Influence of W/C, aggregate type and combination (Series I)

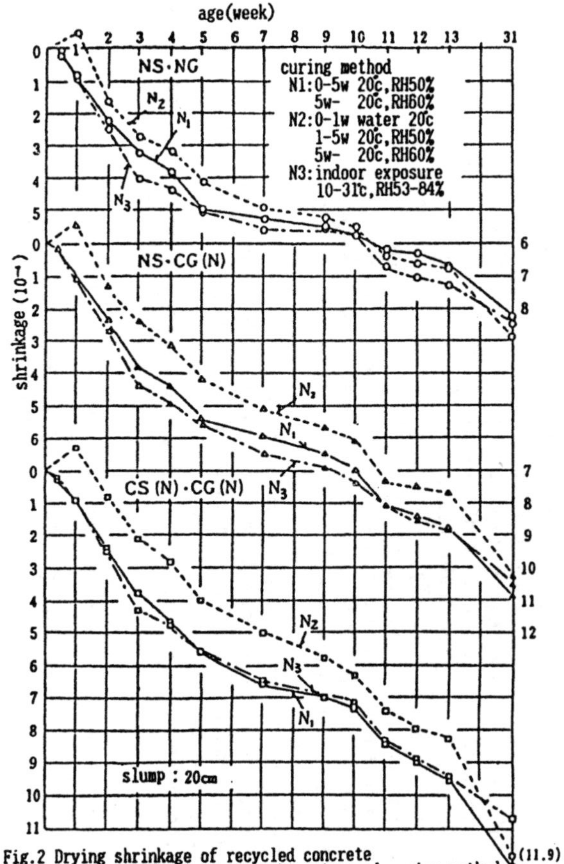

Fig.2 Drying shrinkage of recycled concrete
Influence of aggregate combination and curing method (Series II)

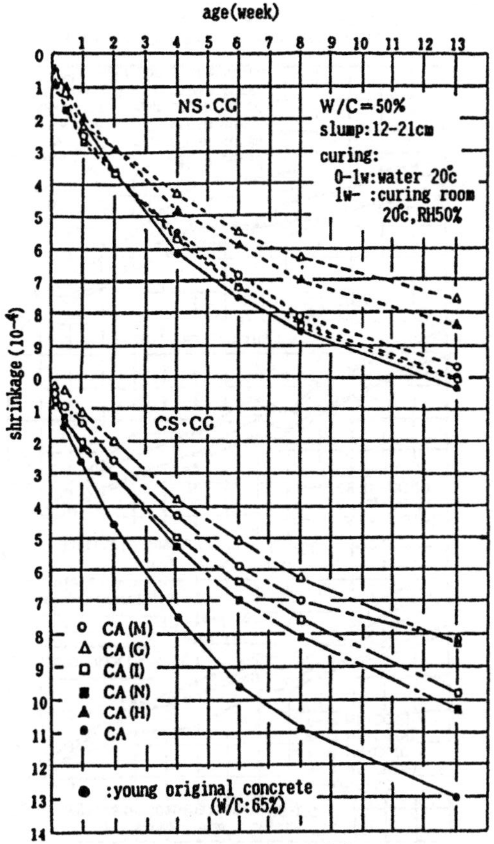

Fig.3 Drying shrinkage of recycled concrete
Influence of aggregate type (Series I)

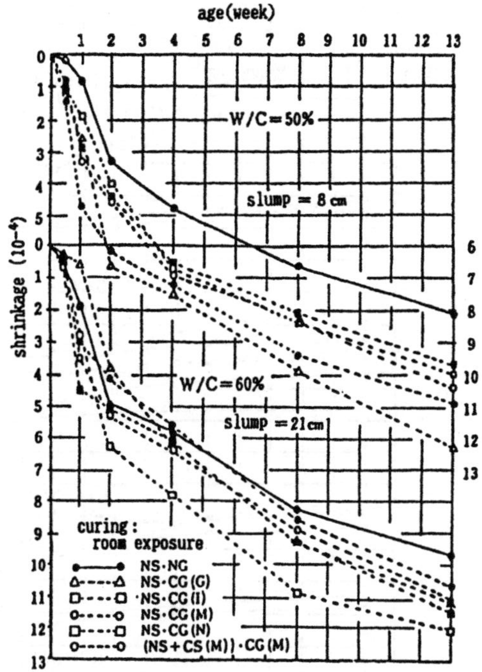

Fig.4 Drying shrinkage of recycled concrete
Influence of W/C and slump (Series III)

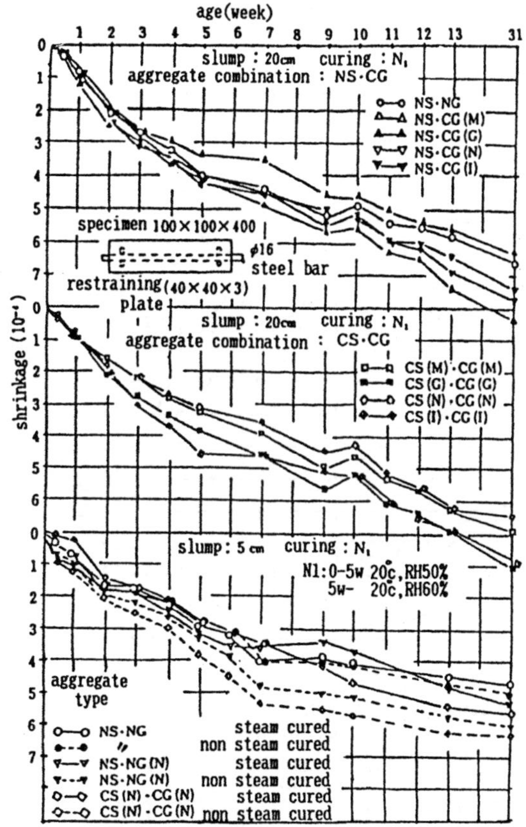

Fig.5 Restrained shrinkage of recycled concrete
Influence of slump and aggregate combination
(Series II)

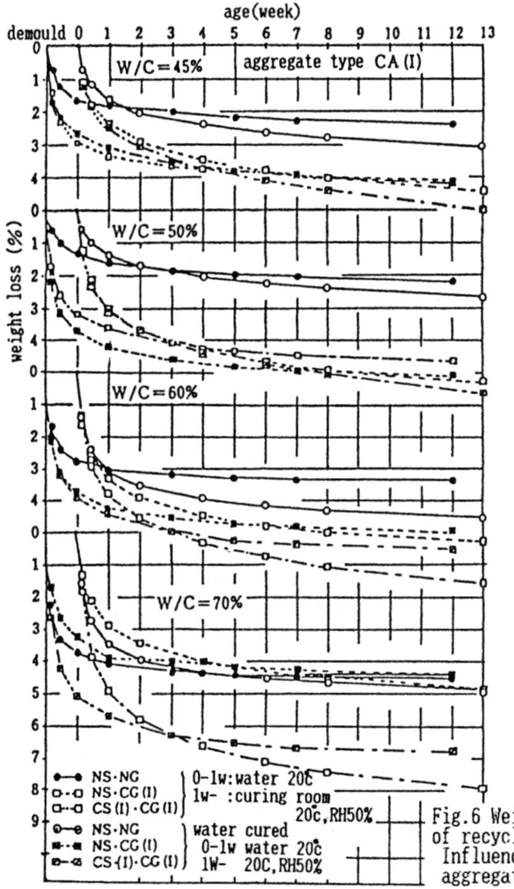

Fig.6 Weight change of recycled aggregate. Influence of W/C and aggregate type

Fig. 3 also shows the past shrinkage result of recycled concrete which used young concrete crushed aggregate. The original concrete of 65% W/C was steam cured for one day and was crushed by jaw crusher at the age of 1 month, and the compressive strength (Fc) was about 250 kgf/cm^2 at crushing. When the age of original concrete is younger, the shrinkage of recycled concrete becomes larger.

This is considered that the cement which adhered on the aggregate surface was less hydrated and caused larger shrinkage.

3) Influence of curing method

Fig. 2 shows the concrete shrinkage at various curing ways, and shrinkage is the smallest by water curing. (Immersed in water for first 1 week after demoulding and indoor exposure afterward)

The curing of curing room exposure under constant temperature (20°C) and relative humidity (50 or 60%) showed similar shrinkage to indoor exposure. (10-31°C, 53-84%)

4) Influence of restraint (See Fig. 5)

In restrained shrinkage specimen, the steel bar of ϕ16mm with anchor plate at both ends was embedded in the concrete specimen to cause the restraint of free shrinkage. Restrained shrinkage was 5-7x10^{-4} at 13 week and 6-8x10^{-4} at 31 week, and smaller than free shrinkage by 30-40%.

Like the case of free shrinkage, restrained shrinkage of NS·NG concrete showed the smallest, and shrinkage of CS·CG concrete is the largest.

3.3 Weight change of recycled concrete (See Fig. 6)

As the W/C and quantity of crushed aggregate used becomes larger, the weight loss of concrete becomes larger. Weight loss of concrete cured by indoor exposure showed larger value than that of concrete cured in water for 1 week and exposed in curing room afterward. The concrete which showed the largest weight loss of 7-8% was CS·CG concrete of 70% W/C, and this concrete is considered to be porous and the water permeation is large. As a result, the quality is considered to be much inferior to NS·NG concrete.

4. Conclusion

Concrete rubble used in this test is considered to be fully hydrated as they had been used for 23 years. Compared to the shrinkage of concrete which used aggregate from young 1 month old concrete, the shrinkage of concrete which used aggregate obtained from 23 years old concrete rubbles was about the half.

In this respect, the concrete crushed aggregate obtained from older concrete is preferable from the viewpoint of shrinkage.

The concrete made by the aggregate combination of NS·CG, CS·CG showed higher shrinkage value by 20-30% than conventional NS·NG concrete, and CS·CG concrete showed larger shrinkage than that of NS·CG concrete. But this shrinkage difference is not so crucial that both NS·CG and CS·CG concrete can be used in practical application.

But when the W/C of CS·CG concrete was large, the weight loss of concrete was also large.

The large weight loss of concrete is considered that the water absorption is large, and care should be paid for their practical application where the freezing-thawing action is strict. From the strength behavior such as compressive strength, tensile strength and Young's modulus, NS·CG and CS·CG concrete showed smaller strength characteristics than NS·NG concrete commonly used.

The strength reduction of NS·CG concrete compared with NS·NG concrete is about 5-10% and that of CS·CG concrete is about 20-30%.

The following conclusions are drawn from above-mentioned points.

(1) NS·CG concrete shows a bit inferior characteristics in strength and shrinkage to NS·NG concrete, but this concrete is considered to be satisfactory in their practical application like NS·NG concrete.

(2) CS·CG concrete shows 20-30% strength reduction and about 20% larger shrinkage than NS·NG concrete. In using this concrete, care should be paid for their environmental condition and importance of structures to be used.

(3) The property of concrete crushed aggregate is usually inferior to natural aggregate due to the detrimental effect of adhered mortar on aggregate surface. Especially crushed fine aggregate is more inferior to crushed coarse aggregate, because they include the powder like micro particles of mortar.

At producing concrete crushed aggregate, the crushing method which can reduce the adhered mortar content on the aggregate surface more would be desirable.

In this respect, the double crushing method which has the secondary crushing after primary crushing to remove adhered mortar on aggregate surface is more recommended.

Acknowledgement

In making this report, the experimental data of Meiji and Nihon University were also used. I will show my thanks to Prof. Takeshi Mukai and Masafumi Kikuchi of Meiji University and Prof. Yoshio Kasai of Nihon University for their offering useful test data.

References

Building Contractors Society (1976) Study on reuse of concrete waste, Part 1 and 2. Committee report, (in Japanese)

Building Contractors Society (1977) Study on reuse of concrete waste. Committee report, (in Japanese)

Fujii, T., Mukai, T., Kikuchi, M., Kasai, Y. and Yamada, T.(1976) Study on concrete which uses concrete crushed aggregate, Part 11 Drying shrinkage. Proceeding of Annual Meeting, Architectural Institute of Japan, pp59-60. (in Japanese)

PROPERTIES OF REINFORCED CONCRETE BEAMS CONTAINING RECYCLED AGGREGATE

TAKESHI MUKAI Engineering Department of Meiji University
MASAFUMI KIKUCHI Engineering Department of Meiji University

Abstract

This investigation was carried out for the purpose of practical application of recycled aggregate concrete as structural concrete.

Reinforeced concrete beams were tested with respect to bending, shearing and bonding under the varied condition of aggregate of different types and method of reinforcement. The results clarified that in case recycled aggregate concrete containing recycled aggregate within 30 %, the properties are nealy same as those of the reinforced concrete beam using sand-gravel concrete.

Key words : Recycled aggregate, Recycled aggregate concrete, Sand-gravel concrete, Reinforced concrete beam, Bending, Shearing, Bonding, Bending creep.

1. Introduction

The amount of waste concrete generated by demolition of reinforced concrete constructions and concrete products thrown out as useless rapidly increased in Japan. The authors, from the viewpoint of effective use of resources, have been continued experimental investigation of the reuse of crushed waste concrete as aggregate for concrete. As a results, it has been proved, as shown in Fig.1 that concrete produced by the recycled aggregate replacing within 30 % or so of sand-gravel has a very meager drop in compressive strength

Based upon the fundamental experiments above-mentioned, we tested bending, shearing and bonding of reinforced concrete beams using recycled aggregate 15 and 30 % replaced of sand-gravel aggregate and checked its applicability in order to aim at the practical use of recycled aggregate.

2. Materials and concrete

2.1 Cement

Ordinary portland cement having compressive strength of 406 kgf/cm at 28th day.

2.2 Aggregate

The types and notations used are shown in Table 1, and physical properties in Table 2.

2.3 Reinforcing bar

Table 3 shows the types and physical properties of reinforcing bar.

2.4 Concrete

Table 4 shows various types of combined fine and coarse aggregate used in our investigation. The concretes used in each test are six types described in Table 5. The concretes of NO.1 through NO.4 were prepared with the compressive strength at 320 ~ 340 kgf/cm², so as to make comparison easy experimented results.

3. Bending test of reinforced concrete beam

3.1 Specimen

Bending strength and bending creep were tested using reinforced concrete beams.The specimens used for bending strength test were of 15× 15 cm cross section and 180 cm long. Six types of specimens each different in kind of concrete and amounts of tension bar as known from Table 6 were used. On the other hand, bendingcreep specimens are 10× 10 cm cross section and 140 cm long,and three kind of specimens were used as shown in Table 6.

3.2 Testing method

Bending strength test was carried out as Fig.2 (a) having span of 150cm, and under the two point concentrated loading method. In the bending creep test, as shown in Fig. 2 (b) two specimen were used as a set, and load was applied at the center of span 120 cm by means of coil spring.Creep deflection was measured for the period of one year. The laboratory was kept at 20℃, RH 60% during the loading period.

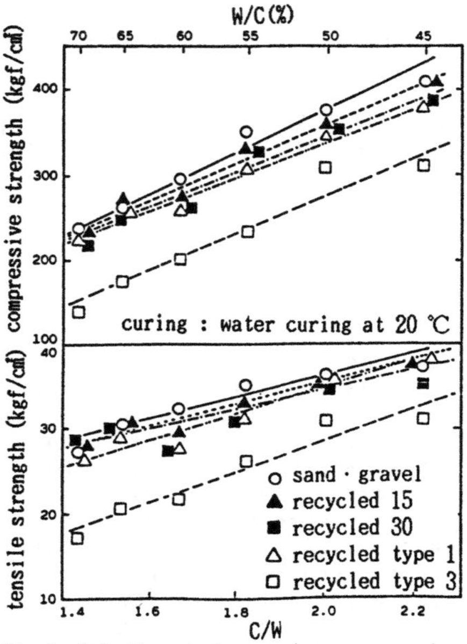

Fig.1 Relations between c/w - compressive strength and tensile strength at 28 days

Table 1 Type of aggregate used in this investigation

Type of aggregate			note
Recycled aggregate	coarse	CG	Crushed by jaw crusher and classified to fine and coarse aggregate by sieve
	fine	CS	
Sand-gravel	coarse	NG	Sand and gravel gathered at the Fuji river
	fine	NS	

Table 2 Physical properties of aggregates used in this investigation

type of aggregate		maximum size (mm)	FM	specific gravity	absorption (%)	unit weight (kg/l)	solid volume percentage (%)
coarse	CG	25	7.19	2.21	6.69	1.23	55.7
	NG	25	6.86	2.63	0.81	1.62	61.7
fine	CG	(5)	3.37	2.06	9.37	1.42	69.1
	NG	(5)	2.91	2.59	0.96	1.75	67.7

Table 3 Physical properties of reinforcing bars used in this investigation

type		yield point stress (kgf/cm²)	tensile strength (kgf/cm²)	Young's modulus (10^6kgf/cm²)	elon-gation (%)
SR-24	4 φ	2560	3560	2.06	45.1
	6 φ	4450	4820	1.92	24.3
SD-30	D13	3620	5690	1.94	23.0
	D16	3800	6010	1.95	25.3
	D19	3960	6210	1.97	24.0

Tble 4 Type of concretes used in this investigation

type of concrete	combination(%) coarse NG	CG	fine NS	CS
sand - gravel	100	0	100	0
recycled 15	85	15	85	15
recycled 30	70	30	70	30
recycled Type 1	0	100	100	0
recycled Type 3	0	100	0	100

Table 5 Mixproportions and physical properties of concretes used in this investigation

NO	type of concrete: clasification	mix proportion: W/C (%)	water content (kg/m³)	S/A (%)	properties at fresh concrete: slump (%)	unit weight (kg/l)	properties at testing: compressive strength (kgf/cm²)	Young's modulus (10^6kgf/cm²)	strain at Pmax (%)	note
1	sand-gravel	60	172	41.5	18.8	2.35	317	3.18	0.21	A[1]
2	sand-gravel	66	185	49.3	18.2	2.27	248	3.27	0.22	B[2]
3	recycled 15	59	176	43.5	18.3	2.30	362	3.08	0.25	A
4	recycled 30	57	180	45.5	19.1	2.25	349	2.90	0.26	A
5	recycled Type 1	62	186	46.1	16.0	2.14	282	2.41	0.25	B
6	recycled Type 3	50	202	49.1	17.6	2.03	298	1.94	0.28	B

1) Concrete mixtures for bending, shearing and bonding test of reinforced concrete beams.

2) Concrete mixtures for bending creep test of reinforced concrete beams.

Table 6 Type and arrengement of reinforced concrete beams for bending and bending creep test (unit:mm)

NO	specimen: notation	main reinforcement: tension side	compression side	stirrup: test zone	non test zone	note
1	A-sand-gravel	2 - D13	2 - 6φ	2-6φ	2-6 φ	
2	A-recycled 15			pitch 100	pitch 40	
3	A-recycled 30	pt:1.4%				for bending
4	B-sand-gravel	3 - D16	2 - 6φ	2-6φ	2-6 φ	test
5	B-recycled 15			pitch 100	pitch 40	
6	B-recycled 30	pt:3.3%				
7	sand-gravel			2-4φ	2-4 φ	for bending
8	recycled Type 1	3 - 6φ	3 - 6φ	pitch 70	pitch 70	creep test
9	recycled Yype 3					

Pt : Percentage of main reinforcement

3.3 Results of experiment and analysis

(1) Bending strength and deflection

(a) Failure mode : No matter what type of concrete was used, A Type specimens (p_t : 1.4 %) failed by the yield of tension bar and B Type specimens (p_t : 3.3 %) demolished by the collapse of concrete at compression side.

(b) Bending strength : In reference to Table 7, the specimen using recycled aggregate was observed the load somewhat lowering when the first crack occured, but ultimate bending strength differed little from the specimen using sand-gravel concrete.

(c) Relative deflection : Fig.3 indicates the relation of load and deflection of each specimems. When deflections at ultimate load are compared, sand-gravel specimen (NO.1) of A Type shows 400×10^{-2}mm, while the specimens using recycled15 and 30 specimen (NO.2, 3) showed the value of 1.5 to 1.8 multiaplication.

In case of B Type specimens, there were little difference.

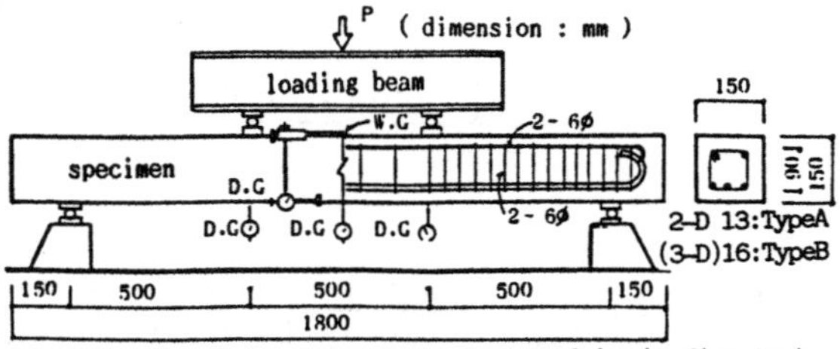

Fig.2(a) Specimen and loading method used for bending test

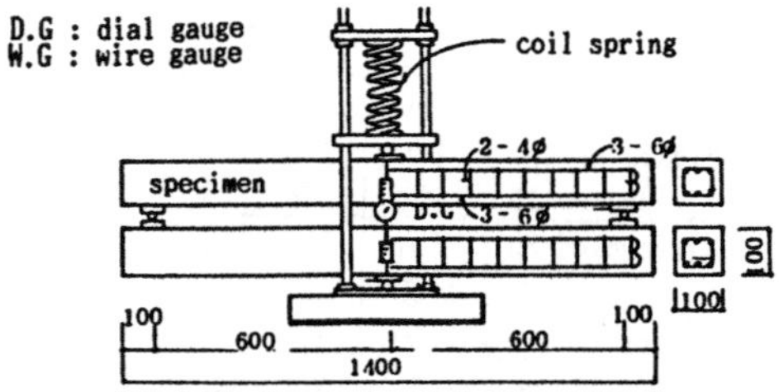

Fig.2(b) Specimen and loading method used for bending creep test

Table 7 Results of bending test

specimen		at first crack occuring		at reinforcement yielding		at ultimate		type of rupture
NO	notation	bending moment (t·cm)	test / calc (1)	bending moment (t·cm)	test / calc (2)	bending moment (t·cm)	test / calc (3)	
1	A-sand-gravel	17.5	0.99	105.0	1.02	111.8	1.06	A
2	A-recycled 15	15.0	0.78	110.3	1.08	125.8	1.19	A
3	A-recycled 30	14.3	0.76	105.0	1.02	125.3	1.19	A
4	B-sand-gravel	12.5	0.71	—	—	215.0	0.91	B
5	B-recycled 15	17.5	0.91	—	—	225.0	0.95	B
6	B-recycled 30	15.0	0.80	—	—	230.0	0.97	B

· Calculated values:These were calculated by formulas given by Architectural Institute of Japan

(1) $Mc=1.8\cdot\sqrt{Fc}\cdot Ze$ (2) $My = a_t\cdot f_y\cdot j$ (3) $M_u = 0.9\cdot a_t\cdot f_y\cdot j$

· Type of rupture

A : Failured due to yielding of reinforcement at tension side.

B : Failured due to crushing of concrete at compressive side.

(2) Bending creep

Bending creep deflection of the concrete using recycled aggregate in the entire aggregate or the whole coarse aggregate showed lager bending creep, compared with the specimen of sand-gavel (See Fig. 4).

Furthermore, it was observed in the sand-gravel concrete specimen bending creep proceeding nearly ended in around 13 weeks loading, whereas, the specimen using recycled aggregate continued bending creep in one year's loading.

4. Shearing test of reinforced concrete beam

4.1 Specimen

The size of specimens were 15 ×25 cm for cross section, and 210 cm long, as shown in Fig. 5 (a). The type of specimens are six shown in Table 8 with varied kind of concrete and amounts of stirrup. The specimens were similary prepared as bending specimens.

4.2 Testing method

For load application, antisymmetric loading method was used as shown in Fig.5 (a).

In this test, each specimen was shearing span 60 cm constant.

Specimen were tested by the regular increasing loading system as shown in Fig. 5 (b).

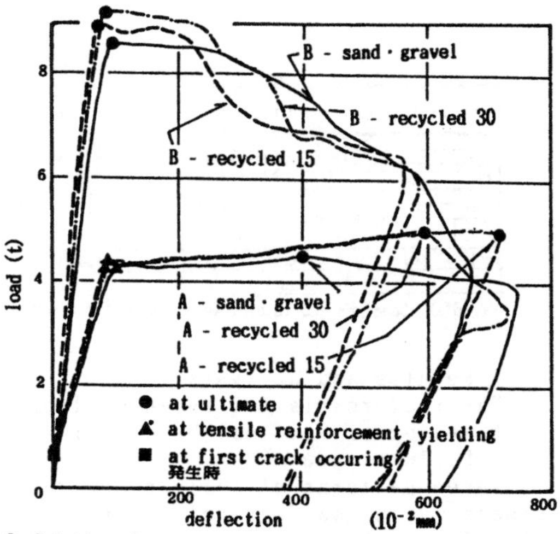

Fig.3 Relations between load and deflection of beam specimens

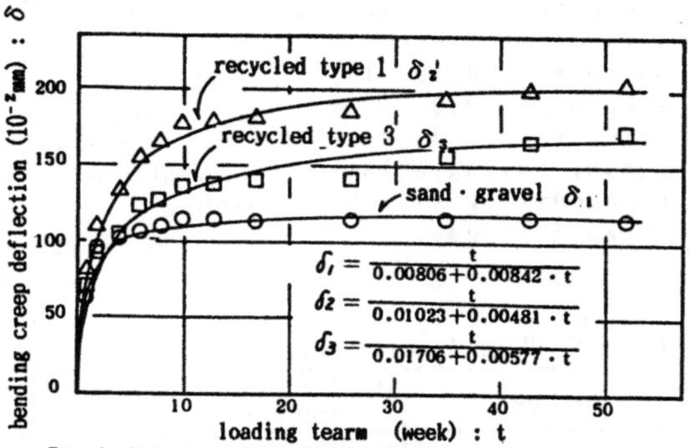

Fig.4 Relations between loading age and bending creep deflection

Table 8 Type and arrengement of reinforced concrete beams for shearing test (unit:mm)

specimen		main reinforcement		stirrup	
		tension	compression	test	non test
NO	notation	side	side	zone	zone
11	A-sand-gravel	3 - D16	3 -D16	2-6φ pitch 100 p_w 0.37%	2-6 φ pitch 40
12	A-recycled 15				
13	A-recycled 30				
14	B-sand-gravel	3 - D16	3 -D16	2-6φ pitch 100 p_w 0.75%	2-6 φ pitch 50
15	B-recycled 15				
16	B-recycled 30				

Pw : Percentage of stirrup

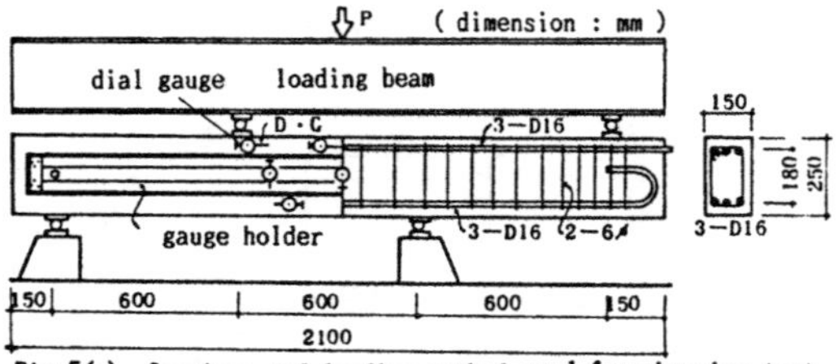

Fig.5(a) Specimen and loading method used for shearing test

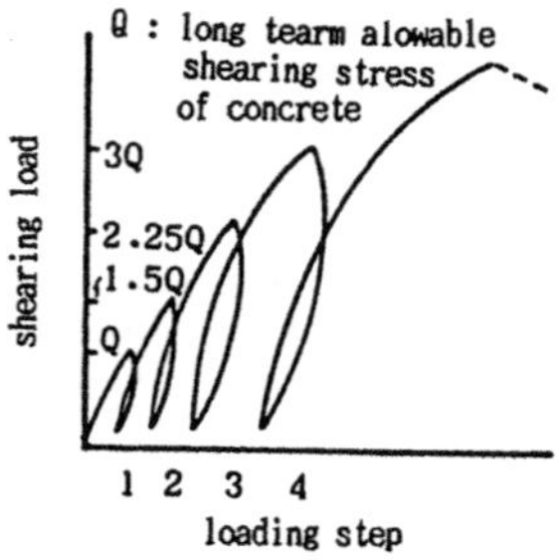

Fig.5(b) Repeated loading method used for shearing test

4.3 Results and analysis

The test resuls are shown in Table 9, Fig.6 and 7. Brief points of these experiment may be described as follows.

(a) Failure mode

Three patterns of failure are revealed in Table 9. NO.11,14 specimens using sand-gravel concrete showed shearing tensile failure (B mode) in Fig. 6. In contrast, NO.12, 13 specimens(Pw : 0.37%) using recycled aggregate, having a meager amount of stirrup showed (A) mode failure, whereas, NO.15, 16 specimens(Pw:0.75 %) using recycled aggregate, having a large amount of stirrup shown (C) mode.

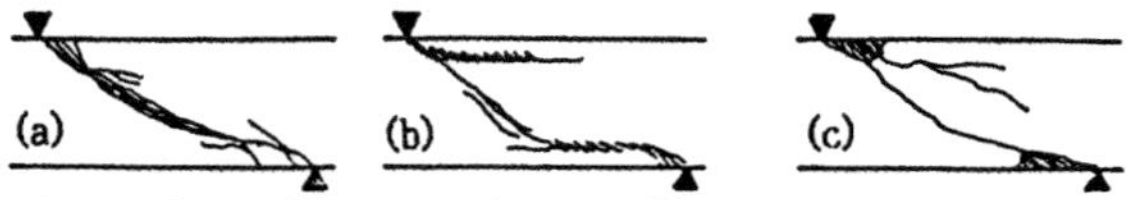

Fig.6 Failure modes of beams specimen occured by shearing test

Table 9 Results of shearing test

specimen		at first crack occuring		at diagonal crack occuring		at ultimate		type of
NO	notation	bending moment (t・cm)	test / calc (1)	shearing stress (kgf/cm²)	test / calc (2)	shearing stress (kgf/cm²)	test / calc (3)	rupture
11	A-sand-gravel	45.0	0.92	18.6	1.02	47.0	1.08	B
12	A-recycled 15	67.5	1.26	20.6	1.05	44.3	0.93	A
13	A-recycled 30	52.5	1.01	17.7	0.92	40.6	0.87	A
14	B-sand-gravel	60.0	1.22	17.7	0.97	56.7	1.17	B
15	B-recycled 15	67.5	1.26	19.5	0.99	63.8	1.22	C
16	B-recycled 30	60.0	1.15	19.5	1.02	61.0	1.19	C

・Calculated values were gotten by following formulas

(1) $M_c = 1.8\sqrt{F_c} \cdot Z_e$ ------------------ (Architectural Institute of Japan)

(2) $\tau_c = K_c(500 + F_c)\dfrac{0.085}{M/Qd + 1.7}$ --------- (Proposed by Ohno and Arakawa)

(3) $\tau_c = K_u \cdot K_p(180 + F_c)\dfrac{0.12}{M/Qd + 0.12} + 2.7 \cdot \sqrt{P_w \cdot {}_s\sigma_y}$ ---------

(Proposed by Ohno and Arakawa)

・Type of rupture : see Fig.7

(b) Shearing strength

The ultimate strength of NO. 12, 13 specimens using recycled aggregate having meager amount of stirrup showed nearly 10 % strength drop compared with NO.11 specimen using sand-gravel. However, NO. 15, 16 specimens having large stirrup revealed shearing strength greater than NO.14 specimen using sand-gravel concrete. This proves that the effect of shearing strength of reinforced concrete beam using recycled aggregate concrete is sufficient. Test values and calculated values in each load stages agreed marvelously.

Accordingly, the calculation formulas can well be applied to reinforced concrete members using recycled aggregate.

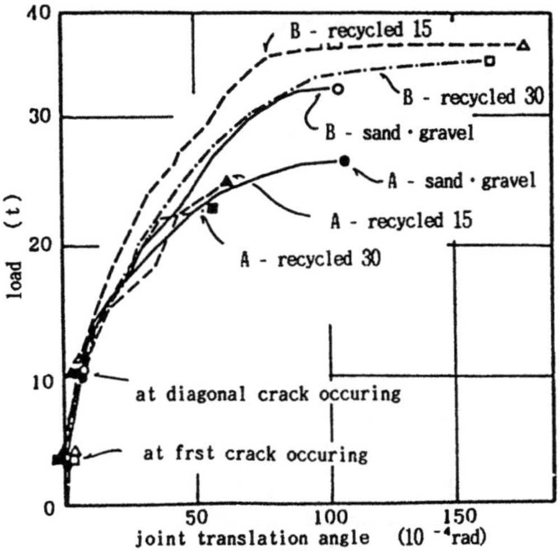

Fig. 7 Relation between shearing load and joint translation angle

(c) Joint translation angle

Fig. 7 showed the relation of the load of each specimen and joint translation angle. Ultimate joint translation angle of the specimen using recycled aggregate concrete 15 and 30 showed weak toughness such as 60×10^{-4} rad, or so in the specimens of a small amount of stirrup. Against it, the specimens with larger amount of stirrup showed nearly 3 times. In this case, the ductility was superior to compare with the specimens using sand-gravel concrete.

5. Bonding test of reinforced concrete beam

5.1 Specimen

The size of specimens for bonding test are shown in Fig. 8, namely, 20× 25 cm cross section, and 78 cm long. Table 10 lists the six types of concrete and stirrup.

The specimens are prepared and cured similarly as bending test specimens.

Table 10 Type and arrengement of reinforced concrete beams for bonding test (unit:mm)

specimen		main reinforcement		stirrup	
NO	notation	tension side	compression side	test zone	non test zone
21	A-sand-gravel	2 - D16	2 -D19	2-6φ pitch 100 p_w 0.22%	2- 6φ pitch 50
22	A-recycled 15				
23	A-recycled 30				
24	B-sand-gravel	2 - D16	2 -D19	2-6φ pitch 100 p_w 0.45%	2- 6φ pitch 50
25	B-recycled 15				
26	B-recycled 30				

Pw : Percentage of stirrup

5.2 Testing method

Load was applied as shown in Fig. 8. The slip of concrete at the outside and free end of bar was measured by means of dial gauge.

Loading was applied as shown in Fig. 9, dividing into ten control stages the amount of slip of free end of bar, while repeating 5 times in the positive direction on the each controled slir stage.

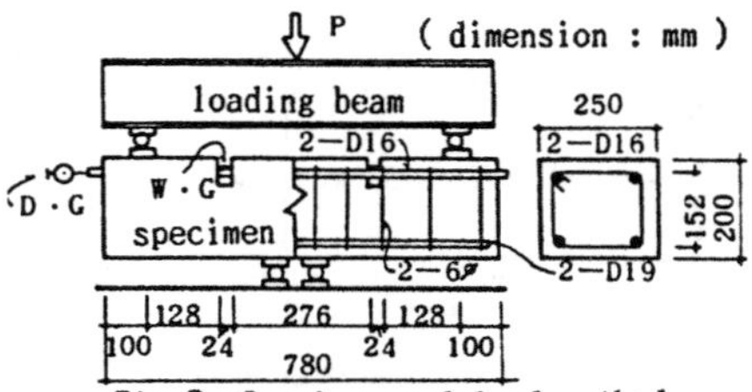

Fig.8 Specimen and load method used for bonding test

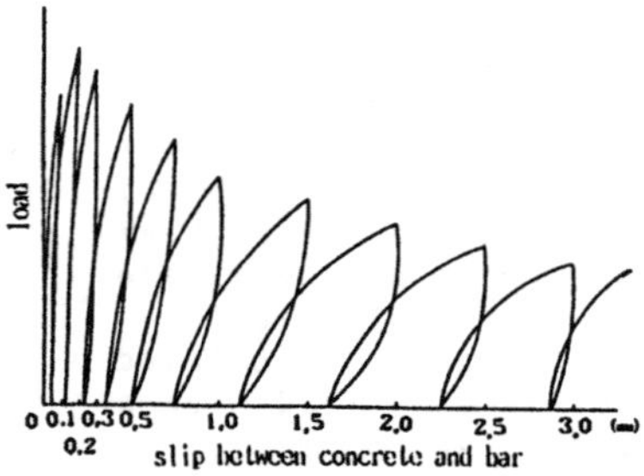

Fig.9 Steps of slip control between concrete and bar of free end

5.3 Results of experiments and analysis

Results are shown in Table 11 and Fig. 10. These are briefed as follows.

(a) Failure mode

Failure was indicated in the case of NO.25 specimen by the yielding of tension bar. But, on all other specimens, ulitimate failure was shown by the split of concrete caused by bonding failure at the position of tension bar.

Table 11 Results of bonding test

specimen		at 0.1mm slip of bar		at bonding crack occuring			at ultimate		
NO	notation	bond stress τ_b 0.1 (kgf/cm²)	$\frac{\tau_b 0.1}{\tau_b max}$	bond stress τ_{bc} (kgf/cm²)	$\frac{\tau_{bc}}{\tau_b max}$	slip of free end of reinforcement (mm)	bond strength τ_b max (kgf/cm²)	$\frac{\tau_b max}{\sqrt{Fc}}$ (1)	slip of free end of reinforcement (mm)
21	A-sand-gravel	58.6	0.90	35.2	0.50	0.02	59.4	3.31	0.08
22	A-recycled 15	57.7	0.89	29.2	0.46	0.02	64.9	3.43	0.75
23	A-recycled 30	63.3	0.98	35.1	0.54	0.03	64.7	3.44	0.05
24	B-sand-gravel	61.5	0.96	35.1	0.55	0.03	64.4	3.58	0.20
25	B-recycled 15	75.6	—	44.4	—	0.09	—	—	0.18
26	B-recycled 30	70.8	0.92	42.4	0.55	0.03	72.2	4.11	0.50

(1) Fc : Compressive strength of concrete

(b) Bonding strength

In reference to Table 11, the $\tau_{bmax}/\sqrt{Fc}$ values of specimens(NO. 22, 23, 26) using recycled aggregate concrete 15 and 30 showed 3.66 in average, which value was nearly 6 % higher, as compared with the specimens(NO.21, 24) using sand-gravel concrete. As for the increasing in bonding strength due to enlarged amount of stirrup, it showed more in comparison with the case of using sand-gravel concrete. Reasoning from such results, the use of recycled aggregate has no problem at all in respect to bonding strength of reinforced concrete beam.

(c) Bonding fatigue

Fig. 10 shows the comparing results each controled slip stage of all specimens between the bonding stress at the first loading and the bonding stress at the fifth repeated loading.

The rate of reduction of bonding stress under the repeated use of load was found out that concrete specimen using sand-gravel concrete and that using recycled aggregate were 31 % and 27 %respectively.

Therefore, we can say that the specimens using recycled aggregate indicate less bonding fatigue between reinforcing bar and concrete than the specimen using sand-gravel.

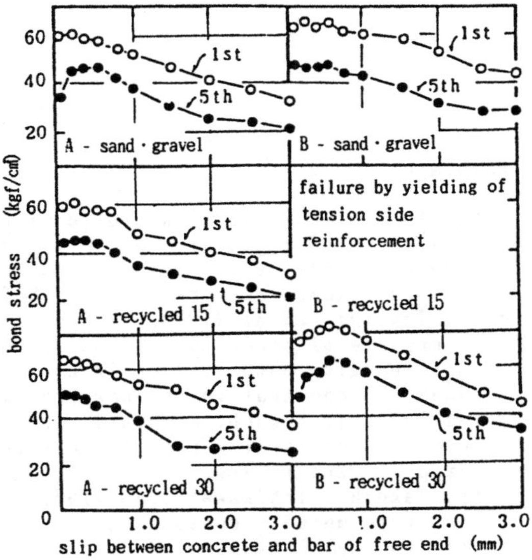

Fig.10 Reduction of bond stress by repeated load (comparison 1st loading and 5th loading at the same slip stage)

6. Conclusions

This investigation was carried out for the purpose of using recycled aggregate for structural concrete. As a results of these experiments, the following facts may be pointed out as conclusions.

(1) Nearly equivalent physical properties were known to exist between sand-gravel concrete and recycled aggregate concrete in respect to bending strength of reinforced concrete beam.

(2) At the point of shearing strength of reinforced concrete with a little amount of stirrup, concrete using recycled aggregate showed slight inferiority, but when the amount of stirrup is enlarged, equal strength can be possessed as the specimen using sand-gravel concrete.

(3) The reduction of bond strength by the use of recycled aggregate has never been observed.
(4) As conclusions, after all these experiments, we can state that little variation in the physical properties of reinforced concrete beam is brought about by the mixture of recycled aggregate with in 30 % in the sand-gravel aggregate.

Accordingly, concrete mixed with recycled aggregate can be used sufficiently as a structural concrete.

Reference

BCSJ (1977) Proposed standard for the " Use of recycled aggregate and recycled aggregate concrete " Building Contractors Society of Japan. Commitee on disposal and reuse of construction waste " May 1977 (English version published in June 1981)

Mukai Takeshi et al (1979) " Study on reuse of waste concrete for aggregate of concrete " Paper presented at a seminar on " energy and resource conservation in concrete technology " Japan-US cooparative science programme, San Francisco.

Mukai Takeshi, Kikuchi Masafumi (1978) " Study on the properties of concrete containing recycled aggregate " Cement Association of Japan " 32nd review.

Mukai Takeshi, Ichikawa Nobuo (1982) " Behaviours of reinforced concrete members containing recycled aggregate(Part I, Behaviors of bending) " Summary of Technical Paper of Annual Meeting. Architectural Institute of Japan.

Mukai Takeshi, Ichikawa Nobuo (1982) " Behaviours of reinforced concrete members containing recycled aggregate(Part II, Behaviors of shesring and bonding) " Summary of Technical Paper of Annual Meeting. Architectural Institute of Japan.

PREPLACED AGGREGATE CONCRETE MADE FROM DEMOLISHED CONCRETE AGGREGATES

TORU KAWAI Institute of Technology
Shimizu Corporation
MORISHIGE WATANABE Nuclear Power Division
Shimizu Corporation
SHIGEYOSHI NAGATAKI Department of Civil Engineering
Tokyo Institute of Technology

Abstract
This paper describes the experiments carried out to determine the physical properties of preplaced aggregate concrete made from demolished concrete of 200mm maximum size.

First, the characteristics of strength were compared between concrete made from demolished concrete aggregate and concrete made from crushed stone.

Second, 40-200mm demolished concrete aggregate was divided into four gradings according to the fragment size and cast into specimen of ϕ600mm × h1200mm cylinder with grout mortar. The compressive strength of concrete made from all these gradings was determined. Enough compressive strength, 15-29MPa, was found for structural concrete.

Third, grouting experiments were conducted with 40-200mm demolished concrete aggregate. Flow gradient of grout mortar and compressive strength of mortar cores were determined.

Finally, effectiveness of preplaced aggregate concrete made from demolished concrete aggregate and applicability to structural concrete were suggested.
Key Words: Preplaced aggregate concrete, Demolished concrete aggregate, Grading, Flow gradient, Grout mortar.

1. Introduction

Two to four million ton of demolished concrete is produced every year in Japan. It is assumed that the volume of concrete waste will increase more and more in the near future. To develop the technology to reuse demolished concrete effectively will be essential in construction industry.

To be sure, to recycle aggregates is a significant way to reuse demolished concrete. Hansen and Narud (1983) published the effective data on strength of recycled concrete made from crushed concrete coarse aggregate. However recycling includes complicated processes such as crushing, selecting, and grading.

To enhance the production efficiency of recycled aggregate, the authors have proposed the method to utilize the larger size of demolished concrete aggregate than the general recycled aggregate as the coarse aggregate for preplaced aggregate concrete.

The authors carried out several experiments to determine the physical properties of preplaced aggregate concrete made from 20-200 mm size demolished concrete aggregate, aiming at the application to man-made rock or other concrete structures.

2. Materials

2.1 Demolished concrete aggregate

Demolished concrete aggregates were prepared from the demolished building structure completed in 1962. Mix proportion of the original concrete is given in Table 1. Compressive strength of the core taken from the 25-year building structure is shown in Table 2.

Demolished concrete was divided into four gradings according to fragment size, 20-40, 40-80, 80-150, and 150-200mm. Physical properties of demolished concrete in each grading and conventional crushed stone are presented in Table 3. From results presented in Table 3 it is apparent that demolished concrete aggregate has much lower density, much higher water absorption, higher loss on ignition, and much higher Los Angeles abrasion loss percentage than crushed stone. Thus, judging from traditional quality indices, demolished concrete aggregate has several inferior characteristics to crushed stone due to considerable amount of original mortar in demolished concrete aggregate.

Table 1. Mix proportions of original concrete.

Specified concrete strength f'c (MPa)		22.1
Maximum size of coarse aggregate (mm)		25
Water cement ratio W/C (%)		56
Sand percentage s/a (%)		41.7
Air content (%)		2.0
Unit content of concrete (kg/m^3)	Water W	158
	Cement C	283
	Fine aggregate S	777
	Coarse aggregate G	1122
Calculated unit weight of concrete (t/m^3)		2.34

C:Ordinary portland cement G:Natural gravel

Table 2. Strength of original concrete.(At the age of 25 years)

Description	Number n	Mean value $\bar{x}$	Coefficient of variation (%)
Compressive strength (MPa)	6	31.3	7.25
Modulus of elasticity (GPa)	6	21.4	12.4
Tensile strength (MPa)	3	2.96	11.0

Table 3. Physical properties of demolished concrete aggregate and crushed stone.

Type of aggregate	Demolished concrete aggregate				Crushed stone
Size fraction of aggregate (mm)	20~40	40~80	80~150	150~200	20~40
Specific gravity	2.39	2.35	2.34	2.34	2.72
Water absorption (%)	7.30	7.51	7.60	7.54	0.76
Percentage of solid volume	59.1	57.1	56.6	57.0	56.5
Loss on ignition (%)	7.88	8.11	8.64	8.04	6.48
Los Angeles abrasion loss percentage (L1000)	32.0	33.4	-	-	23.0
B.S.aggregate crushing value (%)	22.8	-	-	-	20.6
B.S.aggregate 10% crushing load (t)	12.8	-	-	-	17.2

2.2 Grout mortar

Ordinary portland cement was used, similar to ASTM Type I in composition and properties. Fly ash was used as an admixture. River sand (F.M.=1.96) was used as a fine aggregate. Two kinds of chemical admixtures were used. The grout mortar called P-mortar contains the chemical admixture, the main ingredient of which is lignosulfonate.

Table 4. Mix proportions of grout mortar and standard ranges of its properties in the fresh state.

Description			P-mortar		H-mortar	
			P1	P2	H1	H2
Water binder ratio		W/(C+F) (%)	50	48	38	35
Fly ash ratio		F/(C+F) (%)	25	15	10	10
Sand binder ratio		S/(C+F)	1.00	0.990	0.927	0.703
Admixture binder ratio		Ad/(C+F) (%)	1.0	1.0	1.0	1.0
Unit content of mortar (kg/m^3)	Water	W	400	395	351	364
	Cement	C	599	700	831	935
	Fly ash	F	200	124	92	104
	Sand	S	799	815	835	730
	Admixture	Ad	7.99	8.24	9.24	10.4
Efflux value through P-funnel (sec)			16~20		30~40	
Expansion ratio (%)			5~10		2~5	
Bleeding ratio (%)			Less than 3%		Less than 1%	

P-mortar is a representative one for conventional preplaced aggregate concrete. The grout mortar called H-mortar contains the chemical admixture, the main ingredient of which is naphthalene sulfonic acid formaldehyde high condensate. H-mortar is a newly developed one and it has considerable low water-cement ratio within the range, 33-40%. Thus H-mortar is mainly utilized for high strength preplaced aggregate concrete.

Expansion and bleeding ratio of grout mortar were measured with mess cylinders. Flowability of grout mortar was represented in efflux value through P-funnel. Standard ranges of those values are shown in Table 4.

3. Compressive strength

3.1 Small specimen

Small specimen of preplaced aggregate concrete was ϕ150mm × h300 mm cylinder and that of grout mortar was ϕ50mm × h100mm cylinder. They were cast in accordance with the JSCE Standard.

In the first series of experiments, strength of preplaced aggregate concrete with 20-40mm demolished concrete aggregate in air-dry state and the same size crushed stone was experimented with four mix proportions of grout mortar described in Table 4. The moulds were removed at the age of 2 days, and specimens were cured for the specified period of ages. The compressive strength was determined at the age of 7, 28, and 91 days. The result is presented in Fig. 1. Within-batch coefficients of variation were small and varied between 1 and 2 percent. The compressive strength f'c was increased directly proportion to the increase in the compressive strength f'm. Almost all the compressive strength ratios of f'c/f'm were distributed from 0.70 to 1.0, which is the range of conventional preplaced aggregate

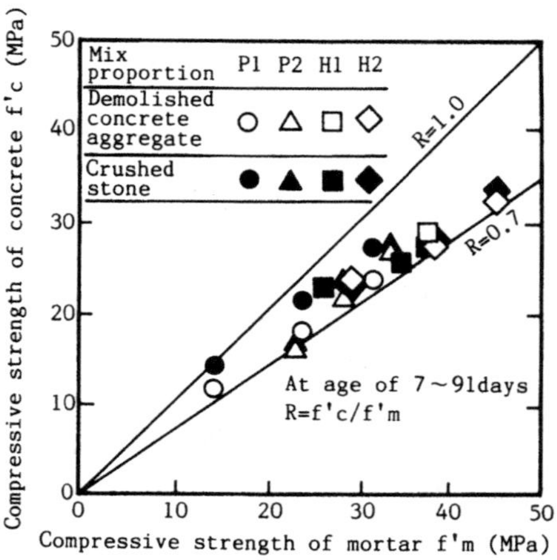

Fig. 1. Comparison of compressive strength.

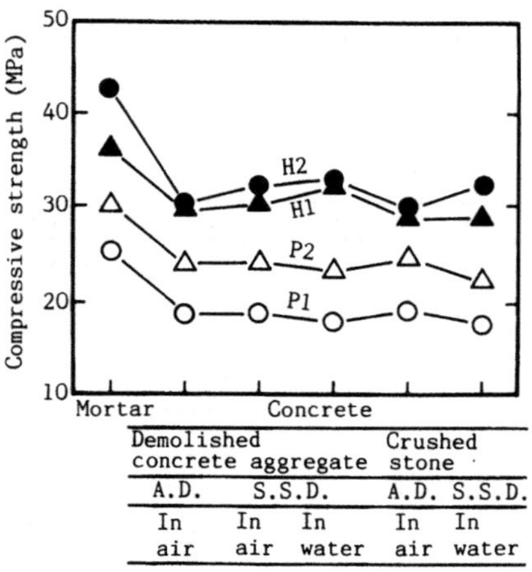

Fig. 2. Compressive strength under different conditions.

concrete. Maximum value of f'c reached approximately 34MPa that exceeded the strength of the original concrete. There were no significant differences in strength of concrete between demolished concrete aggregate and crushed stone.

In the second series of experiments, the small specimens were cast under three different conditions. The first condition was that grout mortar was poured in air-dry (A.D.) state aggregate in air, the second was in saturated surface-dry (S.S.D.) state in air, and the third was in S.S.D. state in water. Four mix proportions of grout mortar were applied, and specimens under three conditions were made from the same batch at the same time in each mix proportion of grout mortar. Result is plotted in Fig. 2. The result indicates that only slight differences occurred among the values.

Data in Fig. 2 show that despite several inferior qualities, demolished concrete aggregate hardly affects the compressive strength of preplaced coarse aggregate concrete under all conditions.

3.2 Large-scale specimen

Compressive strength of preplaced aggregate concrete made from 40-200 mm demolished concrete aggregate was determined. Large-scale specimen was ϕ600mm × h1200mm cylinder. The diameter of specimen was determined to be three times as long as the maximum size of demolished concrete aggregate. According to fragment size, demolished concrete aggregate was divided into four gradings of 40-80, 80-150, 150-200, and 40-200mm.

Mortar was grouted through a grout pipe which was vertically installed at the center of the specimen. The raising speed of the mortar grouting was adjusted to 1.5m/h, considering actual construc-

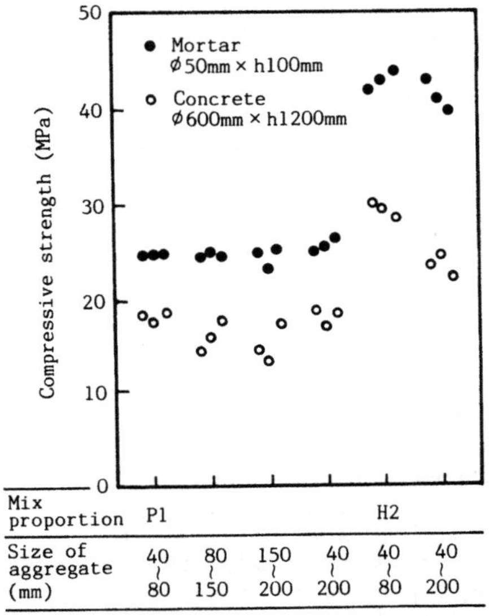

Fig. 3. Compressive strength of large-scale specimen.

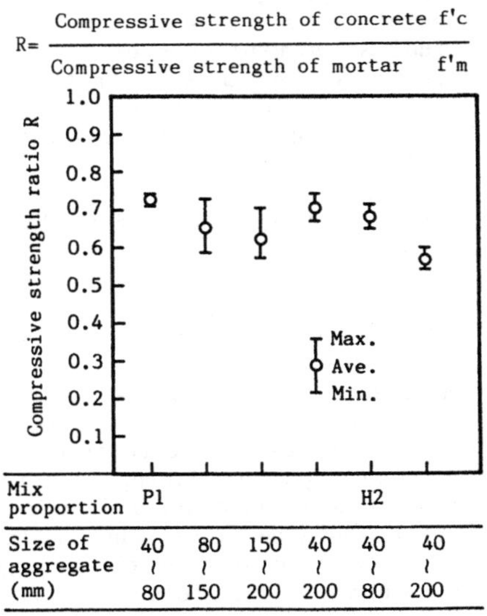

Fig. 4. Compressive strength ratio.

tion. The specified weight was applied on the top of the specimen to confine the mortar expansion immediately after finishing the mortar grouting. The mould was removed in 2 days, then the specimen was cured in water at 20°C until the age of 28 days, and compressive strength and modulus of elasticity were determined. Modulus of elasticity was calculated by the stress and the mean strain of three vertical ones with each 60cm interval at the center of specimen.

Result is presented in Figs. 3 and 4. Comparison of compressive strength ratios of first three different gradings in Fig. 3 indicated the reduction in strength with the increase of maximum size of demolished concrete aggregate when other factors were essentially identical. Size effect on strength occurs because of the difference in size of demolished concrete aggregate. This trend is also explained by Kawakami (1971). He found that a reduction in compressive strength in proportion to log d (d is the size of the coarse aggregate) occurs in the conventional concrete with uni-size of coarse aggregate and with constant volume of aggregate.

Figure 3 shows that compressive strength with demolished concrete aggregate of 200mm maximum size was 15-18MPa in case of P-mortar and 23-29MPa in case of H-mortar respectively. It is apparent from results presented in Figs. 1 and 4 that large-scale specimens have necessary size effect on strength because of the differences in specimen and in demolished concrete aggregate. It is concluded from the data presented in Fig. 3 that preplaced aggregate concrete made from demolished concrete aggregate has great possibilities of application to actual structural concrete in terms of compressive strength.

Figure 5 shows the relationship between compressive strength and modulus of elasticity. The relationship is appreciably close to that of the standard formula of ACI.

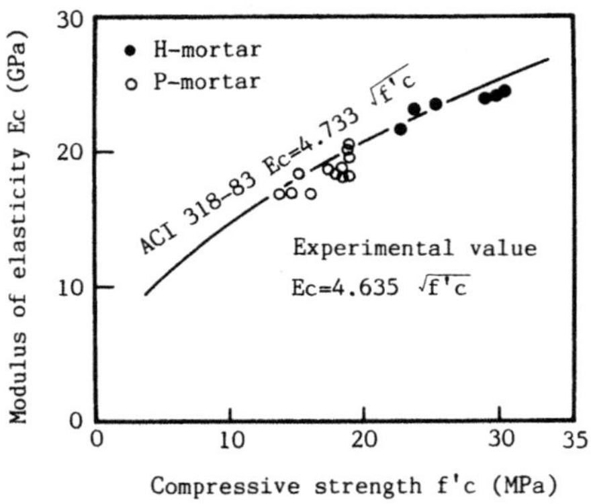

Fig. 5. Modulus of elasticity.

4. Grouting experiment.

Demolished concrete aggregate has more effective water absorption than conventionally used aggregate. Therefore, it seemed that some water in mortar would be absorbed to the air-dry state demolished concrete aggregate during grouting through preplaced demolished concrete aggregate under in-air condition. Gradient of the grout mortar would be higher due to the reduction in flowability of the grout mortar caused by the absorption of water. Taking this assumption into account, following grouting experiments were conducted.

A grout pipe and six pipes with a lot of punched holes 1cm in diameter were placed before placing 40-200mm demolished concrete aggregate within the transparent form as shown in Fig. 6. Mortar was

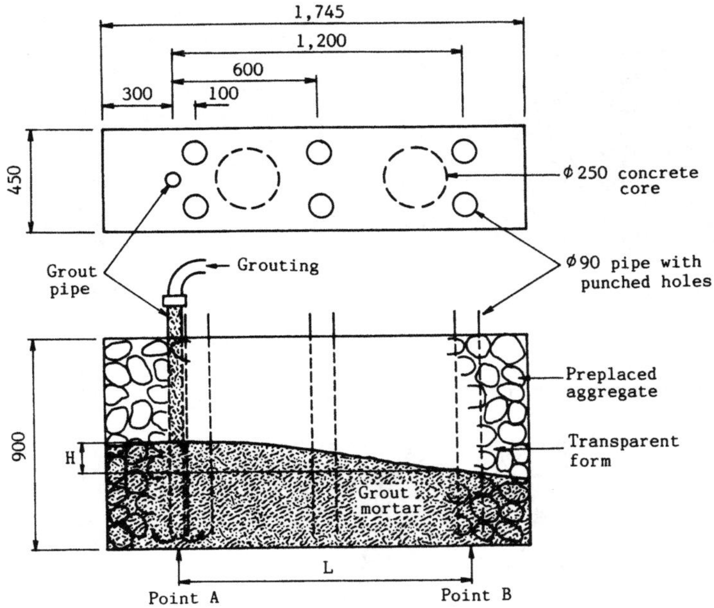

Fig. 6. Outline of grouting experiment.

grouted through a grout pipe. The raising speed of mortar was adjusted to 1.5m/h. Mortar was grouted with two mix proportions of grout mortar P1 and H2 under three different conditions described in Table 5.

Surface height of grout mortar during grouting was measured in every five minutes. Photographs 1 and 2 show the change in gradient of the grout mortar. Figure 7 shows gradients of the grout mortar. They were calculated from the height difference H and the distance L between point A and point B as described in Fig. 6. When using P-mortar gradients showed 1/4 to 1/10. On the other hand, when using H-mortar gradients showed about the half of those of P-mortar. This can be explained by Nagataki et al (1979). They verified that H-mortar was close to a Newtonian fluid. That is, H-mortar has a very

Table 5. Grouting conditions.

No.	Grout mortar	State of moisture in aggregate	Condition
1	P-mortar	Air-dry	In air
2		Saturated surface-dry	In air
3		Saturated surface-dry	In water
4	H-mortar	Air-dry	In air
5		Saturated surface-dry	In air
6		Saturated surface-dry	In water

Photograph 1. Gradient of mortar.

Photograph 2. Gradient of mortar.

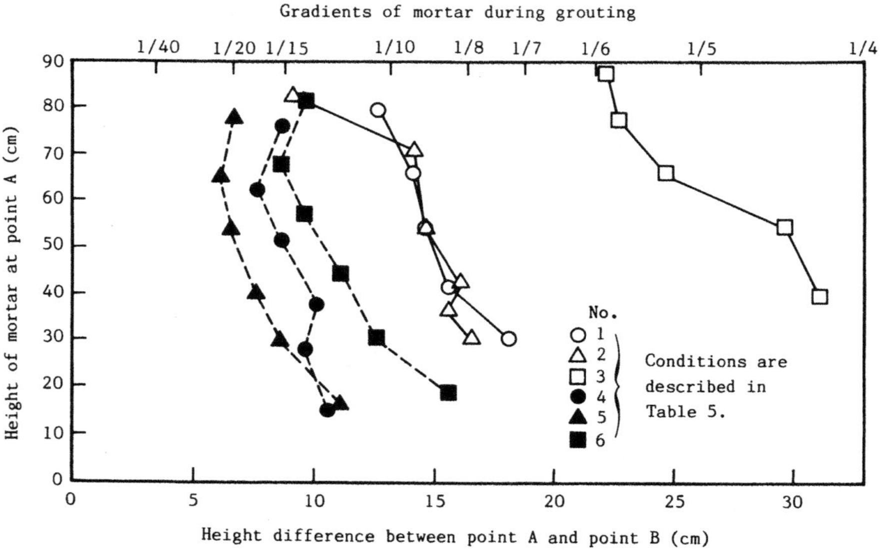

Fig. 7. Gradients of mortar during grouting.

small yield value in close vicinity to zero even if its efflux value through P-funnel is approximately twice as much as that of P-mortar. Figure 7 also shows that gradients under water grouting condition were higher than those in other conditions irrespective of the kind of mortar. These phenomena can be explained by the fact that the water head pressure on mortar reduces the total head pressure difference between given two points.

Almost no significant differences were found in gradient when P-mortar was grouted under in-air condition regardless of the state of moisture in demolished concrete aggregate. On the other hand, differences in gradient occurred according to the state of moisture in demolished concrete aggregate when H-mortar was grouted under in-air condition. However, as mentioned above the gradients under in-air condition were lower than those under in-water condition. Within the scope of these experiments, the absorption of water to

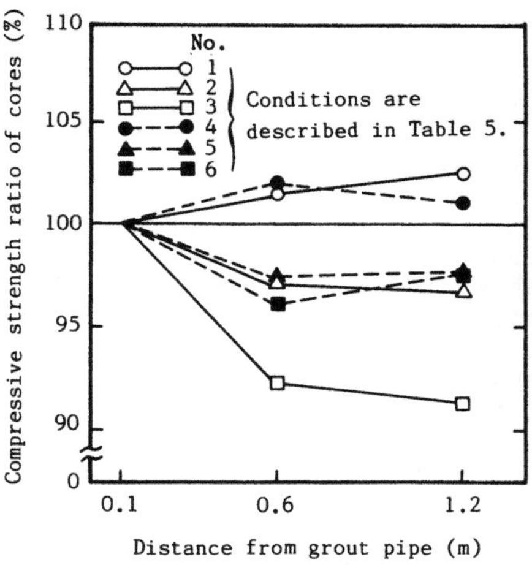

Fig. 8. Change in compressive strength of cores.

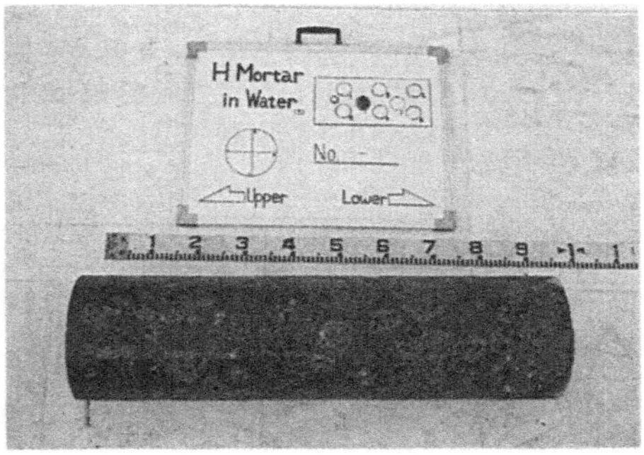

Photograph 3. Appearance of concrete core

the air-dry state demolished concrete aggregate had a slight effect on change in gradient.

Figure 8 shows the relationship between the compressive strength ratio of the drilled core specimens of ϕ55mm × h110mm and the horizontal distance from the grout pipe. The cores were taken from the hardened mortar within the pipe from top to bottom. Core specimens were prepared by cutting every 11cm height. It is indicated from Fig. 8 that compressive strength relatively decreased as the distance increased only when P-mortar was grouted under in-water condition. These data show that the slight segregation occurred. It seems that segregation occurs in grout mortar when it falls through water after it flows over the aggregate. That is, the higher the gradient becomes, the more the segregation occurs. On the other hand, compressive strength slightly increased as the distance increased under air-dry state in-air condition. It suggests that water-cement ratio was gradually reduced through the absorption of water to demolished concrete aggregate.

Taking the results from grouting experiments into consideration, it is concluded that absorbed water to air-dry state demolished concrete aggregate has only a little influence on the change in flowability, and it is also concluded that considerably longer space intervals between grout pipes are applicable when mortar is grouted under in-air condition because the gradient of the grout mortar is lower than that of under in-water condition.

Two concrete cores 25cm in diameter were taken from each specimen to examine the degree of grout filling. Photograph 3 shows the completion of filling.

5. Conclusions

As a result of a series of experiments, we are convinced of the possibility of reuse of demolished concrete as aggregate for preplaced aggregate concrete.

The traditional quality indices for aggregate show that demolished concrete aggregate is inferior to crushed stone. However, there were no significant differences in compressive strength of preplaced aggregate concrete between the one made from demolished concrete aggregate and the one made from crushed stone in case of small specimen. The maximum strength of specimen slightly exceeded the strength of the original concrete.

Size effect on compressive strength occurs in preplaced aggregate concrete due to the difference in size of specimen and due to the difference in maximum size of demolished concrete aggregate. Therefore compressive strengths of the large-scale specimens made from 40-200mm demolished concrete aggregate are smaller than those of small specimens. However, compressive strength of even large-scale specimen using P-mortar and H-mortar spread 15-18MPa and 23-29MPa respectively. P-mortar proved to be applicable to relative low strength concrete structure such as man-made rock. H-mortar proved to be applicable to many types of concrete structures.

6. Future considerations

To establish the preplaced aggregate concrete method which utilizes demolished concrete aggregate, to determine long-time performances such as drying shrinkage and compression creep is indispensable. They are now under experiment.

Furthermore, we are investigating the field experiment to verify the results obtained from this research.

References

Hansen, T. C. and Narud, H. (1983) Strength of recycled concrete made from crushed concrete coarse aggregate. Concrete International, 5, 1, pp79-83.

Kawakami, H. (1971) Effect of coarse aggregate on strength of concrete. Doctoral Thesis, University of Kyoto.

Nagataki, S., Kodama, K. and Okumura, T. (1979) Application of high-strength prepacked concrete to offshore structures. International Symposium on Offshore Structures, RILEM-FIP-CEB Symposium.

PROPERTIES OF RECYCLED AGGREGATE FROM CONCRETE

H.KAGA Technology Planning Department
Taisei Corporation
Y.KASAI Department of Architecture and Architectural Engineering
Nihon University
K.TAKEDA Technology Research Center
Taisei Corporation
T.KEMI....Institute of Technology
Toda Construction

Abstract
Properties of recycled aggregate from concrete were examined. Three different concretes with water cement ratios of 0.45, 0.55, and 0.68 were produced, and they were crushed with a jaw crusher so that aggregates whose maximum size was 25 mm were obtained. The crushed concrete was divided into fine and coarse aggregates, and basic properties of the aggregates were examined. The amount of cement paste in the aggregates was tested for each grading of the aggregates. The results showed that properties of the recycled concrete aggregate were not influenced by the strength of the source concrete, and their properties could not be better than the properties of the original aggregate because the cement paste remained on the recycled concrete aggregate. When the recycled concrete aggregate was stirred in a mixer, better quality of aggregate could be got.
Key words: Reuse of concrete, Crushing of concrete, Recycled aggregates, Properties of aggregates

1. Introduction

Every year more than 3 million cubic meters of concrete from concrete structures are demolished, but as the regulations for disposal become stricter, it is difficult to find places for disposal of waste concrete. It is mainly used as earth fill materials. Every industry tries to reduce the amount of disposal, and reuse of waste concrete should be desirable. It is expected that a large amount of sound material can be obtained from waste concrete.

"A committee of disposal and reuse of wastes from construction works," was established in the Building Contractors Society, Japan, and a study for reuse of waste concrete as aggregate was conducted. This paper is a part of the report of the study, and properties of recycled aggregate from concrete are described in this paper.

2. Production of source concrete

It is considered that concrete from different sources will be used for production of aggregate. In this study, three water cement ratios, 0.45, 0.55, and 0.68 were employed to examine the influence of strength of concrete on characteristics of aggregate produced from the concrete. Materials, mix proportion, and strength of the concrete are shown below.

Materials: Commercially available normal Portland cement, and sand and gravel from the Sagami River were used. The maximum size of sand is 5 mm and that of gravel is 25 mm. Properties of the aggregate are shown in Table 1.

Mix proportion: Mix proportions of concrete are shown in Table 2. Slump of concrete was designed to be 18 cm, and cement contents of concrete with water cement ratio of 0.45, 0.55, and 0.68 are 441, 350, and 280 kg/m3 respectively. The concrete was placed in 15x30 cm cylinder forms or 20x20x20 cm cubic forms so that they were crushed with crushing equipment easily.

Strength of concrete: After steam curing, the concrete was cured in the open air for a month and crushed. Table 2 also shows strength of the concrete.

Table 1. Properties of aggregate for the source concrete

		Gravel	Sand
Specific gravity of s.s.d. condition		2.67	2.65
Content of absorbed moisture (%)		0.8	1.24
Fineness modulus		6.55	3.17
Sieve analysis	30 mm (%)	100	--
	25 mm	98	--
	20 mm	86	--
	15 mm	77	--
	10 mm	50	100
	5 mm	9	98
	2.5 mm	--	77
	1.2 mm	--	56
	0.6 mm	--	34
	0.3 mm	--	14
	0.15mm	--	4

Table 2. Mix proportion and strength of the source concrete

Water cement ratio	0.45	0.55	0.68
Slump (cm)	18	18	18
Sand percentage	38.8	42.3	47.0
Water content per unit volume (l/m^3)	198	193	190
Absolute volume of cement (l/m^3)	140	111	89
sand (l/m^3)	253	289	334
gravel (l/m^3)	399	398	377
Weight of cement (kg/m^3)	441	350	280
sand (kg/m^3)	669	765	884
gravel(kg/m^3)	1065	1063	1007
Compressive strength at 28 days (kg/cm^2)	353	--	231
91 days (kg/cm^2)	385	334	295
crush (kg/cm^2)	375	289	220

3. Crush of concrete

Concrete can be crushed with various equipment, but in this study, a crusher of single toggle swing jaw non-choking type (HA-5008, 45Kw) was employed. In the preliminary crush, the jaw crusher was set at an opening of 25, 35, 45, and 55mm with the jaws in a closed position. Results are shown in Fig.1, and when the open sets were 35 mm, 45 mm, or 55 mm, aggregates of almost the same fineness were produced. Finer aggregate was obtained from the open set of 25 mm.

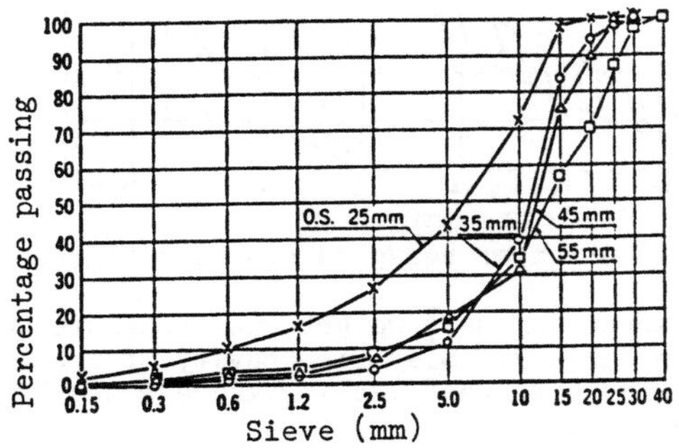

Fig.1 Gradings of recycled concrete aggregate with various open sets

Seventy five to hundred percent of aggregate which was produced from rock with a crusher would go through a sieve whose size is equal to the open set of the crusher, but all aggregate produced from concrete would go through that sieve. When an open set is large, original coarse aggregate in the concrete will pass through a crusher, but when it is small, some particles of the coarse aggregate will be broken. These results indicated that recycled aggregate from concrete which met the standard of 2505 crushed stone in JIS A5005, "Crushed stone for concrete," could be produced when an open set of a jaw crusher was 33 mm. In this study, concrete was crushed by a crusher whose open set was 33 mm.

4. Properties of recycled concrete aggregate

Since about 20 percent of the aggregate from the crusher was under 5 mm, the aggregate was divided by a 5 mm sieve into fine and coarse aggregates, and each aggregate was tested. Some tests were conducted on aggregates of different fineness which were divided by standard sieves.

4.1 Shape and grading

Shapes of recycled concrete aggregate depend on the type of crushers. Aggregate from a jaw crusher is harsh and mortar or cement paste remains on the original aggregate. Percentage of absolute volume of the aggregate was 56 - 58 %, which met the requirement of JIS (more than 55 %) for evaluation of shape.

Particle size distributions of the aggregates under and over 5 mm are shown in Fig.2. The particle size distribution of the coarse aggregate meets the requirements of JIS 2505 crushed stone, but the particle size distribution of the fine aggregate is coarser than the requirements of JIS crushed sand, because the amount of particles between 2.5 and 5 mm are relatively large.

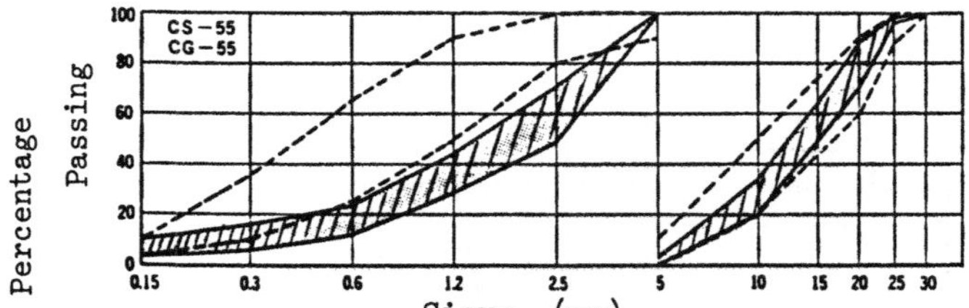

Fig.2 Particle size distribution of recycled concrete aggregate

4.2 Specific gravity, absorption, unit weight, and absolute volume
Test results are shown in Table 3. Specific gravity of absolute dry condition of the coarse aggregate was 2.26 - 2.47 (average 2.34), and that of the fine aggregate was 2.02 - 2.25 (average 2.10). Specific gravity of the aggregates did not change with variation of the source concrete. The specific gravity of the recycled concrete aggregate is lower than that of the original aggregate for production of the source concrete. Specific gravity of the coarse aggregate is near to that of the source concrete, and specific gravity of the fine aggregate is near to that of the mortar in the source concrete. According to JIS, specific gravity of fine and coarse aggregates should be over 2.5, and these results showed that the recycled concrete aggregate had lower specific gravity than the requirements.

Table 3. Properties of the recycled concrete aggregates.

Type of aggregates	Coarse aggregate			Fine aggregate		
Symbols	CG45	CG55	CG68	CS45	CS55	CS68
Fineness modulus	7.06	7.03	7.00	3.81	3.77	3.70
Specific gravity of a.d.condition	2.35	2.33	2.33	2.10	2.10	2.10
Content of absorbed moisture (%)	5.82	6.25	6.40	11.0	11.3	11.1
Unit weight (kg/m3)	1350	1390	1390	1320	1340	1340
Percentage of absolute volume (%)	56.7	57.8	57.9	60.5	61.1	61.2

Content of absorbed moisture of the coarse aggregate was 4.85 - 10.4 % (average 6.46 %), and that of the fine aggregate was 9.80 - 11.6 % (average 11.2 %). When the water cement ratio of the source concrete is high, content of absorbed moisture of the recycled concrete aggregate is also high. Fig.3 shows the absorption of water along time when the aggregate is soaked into water. After 24 hours, the aggregate continues to absorb water. Content of absorbed moisture of the recycled concrete aggregate cannot meet JIS (less than 3 %).

Unit weight of the aggregate was 1.2 - 1.4 kg/l and the coarse aggregate was heavier than the fine aggregate, but percentage of absolute volume of the fine aggregate is 54 - 56 %, and it is higher than that of the coarse aggregate. When the aggregate was packed without jigging, the unit weight was 10 % lower than that with jigging. When aggregate was packed with enough tamping, the unit weight increased by only few percent, which showed that the aggregate was not broken by tamping.

These properties of the aggregate with each fineness are shown in Fig.4. From this figure, it can be said that finer aggregate has lower specific gravity and higher absorption. More details will be discussed in 4.4.

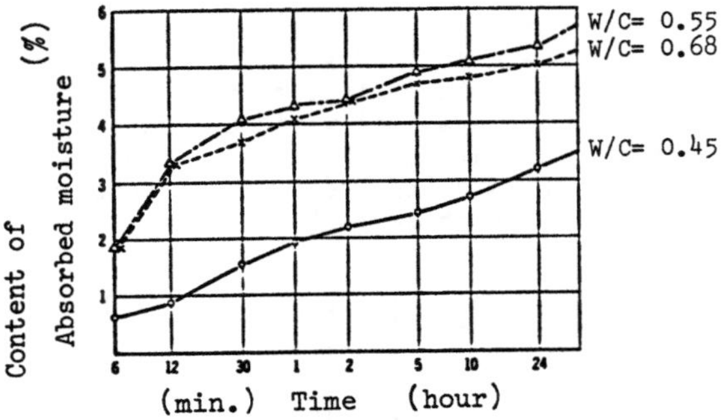

Fig.3 Absorption of water along time

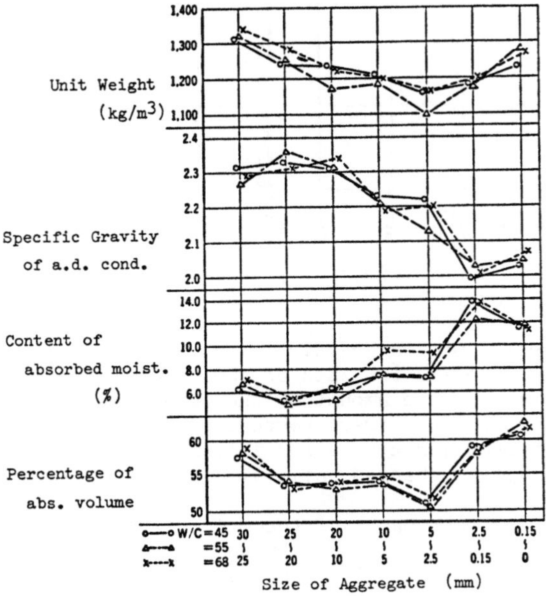

Fig.4 Properties of the recycled concrete aggregates

4.3 Crushing value, soundness, and abrasion
Strength of the aggregate was tested by the crushing test in BS-812. The crushing value was 25 - 36 %, and it was considered that the recycled concrete aggregate was equal or stronger than artificial light weight aggregate or hard light weight aggregate from volcanos. Under higher pressure, mortar in the recycled concrete aggregate is broken, but it is difficult to break the remained original aggregate.

Soundness tests with solution of sodium sulfate showed scattered results. After 5 cycles, about 50 % of the coarse aggregate and about 20 % of the fine aggregate were lost. These results did not meet the requirement in JIS for crushed stone (less than 12 %). These results could be expected because the absorption of the recycled concrete aggregate was high. Through abrasion tests with Los Angels machine, the aggregate lost 25 % of its weight, which met the JIS requirements for crushed stone (less than 40 %). The results of the abrasion tests of the aggregate of various fineness are shown in Fig.5. Coarse aggregate lost its weight more. Almost all weight loss occurred in first 10 minutes of the tests. Aggregate with smooth shape was obtained through these tests.

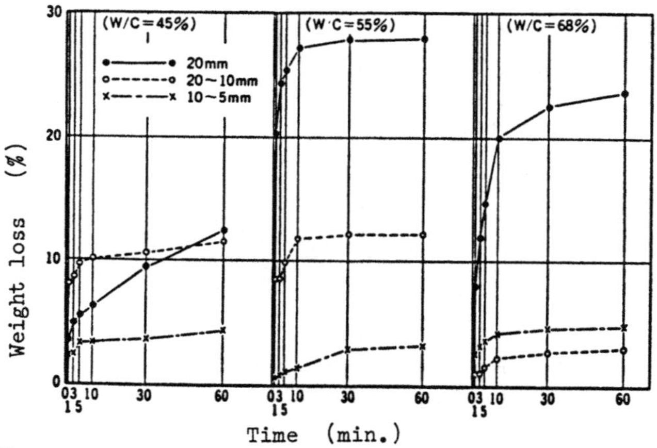

Fig.5 Abrasion tests

4.4 Amount of cement paste with aggregate
Aggregate from crushed concrete consists of the original aggregates and bonded cement paste, and the amount of the bonded cement paste in the aggregate of various fineness were tested. Specimens were soaked into 5 % solution of hydrochloric acid for a long time. The solution was stirred occasionally and changed until all cement paste dissolved away.

It was considered that some parts of the original aggregate would dissolve, but the results were expressed by weight loss without adjustment of the dissolved original aggregate. After the soakage, sieve analysis was conducted. The results are shown in Fig.6 and Table 4.

Twenty percent of the recycled concrete aggregate over 20 mm is cement paste and the finer aggregate has larger percentages of cement paste. Almost half of the aggregate under 0.3 mm is made of cement paste. Table 4 indicates that the particles of the recycled concrete aggregate consist of the original aggregate with various fineness, and larger particles have coarser original aggregate. It is considered that the differences of the specific gravity or absorption of the recycled concrete aggregate with the change of its fineness would come from the differences of the amount of cement paste in the aggregates.

5. Improvement of the aggregate by secondary crush

As the results of the abrasion tests indicated, properties of the recycled concrete aggregate would be improved by secondary crush, which made the surface of the aggregate smooth. Properties of the recycled concrete aggregate after a few-minute stirring in a mixer of a forced mixing type or a tilting mixer were tested.

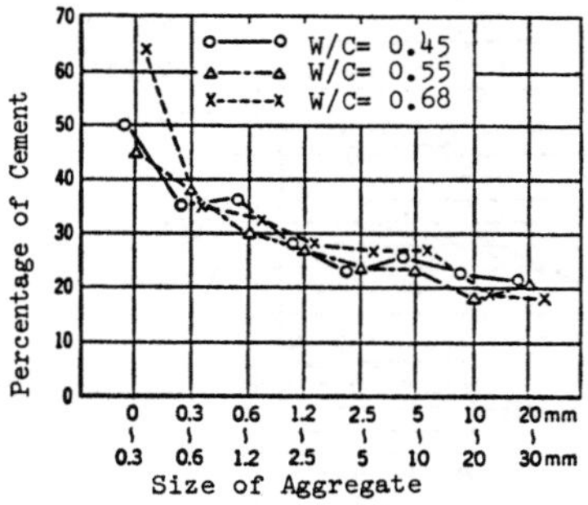

Fig.6 Amount of cement paste with aggregate

Table 4. Grading after soakage in acid

Size	5-10mm(500g)			10-20mm(500g)			20-mm(1,000g)		
Type	CG45	CG55	CG68	CG45	CG55	CG68	CG45	CG55	CG68
Sieve(mm)									
-0.15	0.3	1.1	0.3	0.0	0.5	0.2	0.2	0.2	0.1
.15-0.3	2.1	3.1	3.2	1.5	2.7	1.7	1.8	2.5	1.2
0.3-0.6	4.9	6.5	7.1	3.9	4.7	4.2	3.8	5.4	4.1
0.6-1.2	6.8	8.6	9.7	5.4	6.9	5.9	5.3	6.9	6.0
1.2-2.5	8.4	11.0	12.3	6.4	7.3	7.4	6.3	10.2	7.4
2.5- 5	21.5	26.1	29.2	9.0	11.0	9.4	8.1	12.0	9.4
5 -10	56.0	43.7	37.2	27.1	18.4	17.8	10.5	13.5	10.3
10 -15	--	--	--	37.7	41.6	37.4	17.5	13.5	14.2
15 -20	--	--	--	9.0	6.8	16.0	15.0	14.8	23.3
20 -	--	--	--	--	--	--	31.5	21.0	24.0
Cement(%)	25.6	23.4	27.4	22.2	18.2	19.0	21.3	20.9	18.7

Since properties of the source concrete have little influence on the results, tendency in the whole results is discussed below.

Longer stirring time brought higher percentage of particles under 5 mm, and percentage of the aggregate under 5 mm was about 10 after 5 minute stirring. Weak and angular parts of the aggregate were removed during stirring, and unit weight and percentage of absolute volume of the aggregate increased. Specific gravity of absolute dry condition also increased slightly, and content of absorbed moisture showed 10 - 20 % lower value. Results of BS aggregate crushing value and the abrasion tests showed slightly better value. When the aggregate was stirred over 5 minutes, the shape of the aggregate became more round, but large change of properties of the aggregate could not be seen.

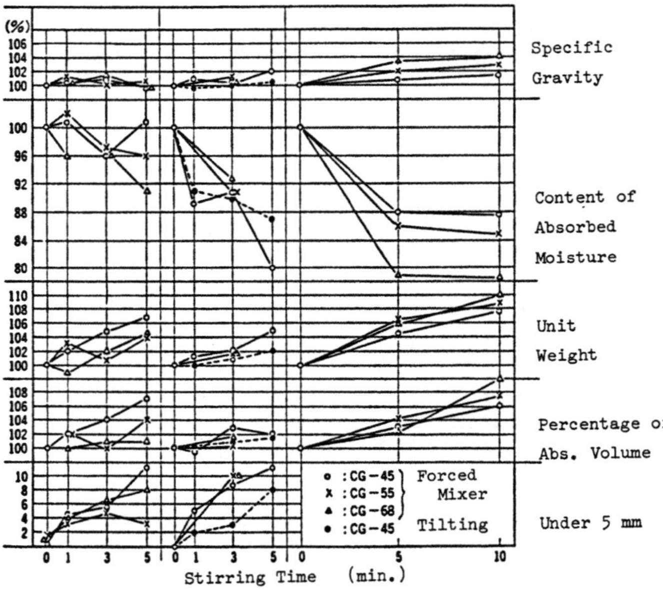

Fig.7 Properties after the secondary crush

6. Conclusion

The purpose of this study is reuse of concrete waste. It is considered that crushed concrete waste can be used for concrete aggregate, and large amount of the concrete waste are expected from demolished buildings.

If all cement paste in concrete can be removed, original aggregate will be recovered and reused. Properties of this aggregate should be equal to old good aggregate. But complete removal of cement paste from the aggregate will need long time and large cost, and it is desirable that crushed concrete can be reused for aggregate without removal of the cement paste.

The recycled concrete aggregate consists of the original aggregate and cement paste, and it was considered that its properties should be influenced by the water cement ratio of the source concrete. In this study, three different water cement ratios, 0.45, 0.55, and 0,68 were employed to produce the source concrete, and the concrete were crushed with a jaw crusher and tested. But the results showed that the strength of the source concrete had little influence on the properties of the recycled concrete aggregate, and similar aggregate could be obtained from different concrete. This is advantage for reuse of concrete waste because the recycled concrete aggregate can be produced without special consideration about variety of strength of source concrete.

When recycled concrete aggregate whose maximum size is 25 mm is produced, 80 % of the aggregate will be coarse aggregate (over 5 mm), and 20 % of the aggregate will be fine aggregate (under 5 mm). But the fine aggregate has too much coarse particles (2.5-5 mm), and it is difficult to use solely as fine aggregate. Properties of the aggregate vary with the size of the particles. Finer particles have lower specific gravity and larger content of absorbed moisture. Difference in percentage of cement paste in the aggregate brings such results, and finer parts of the aggregate have larger percentage of cement paste and poorer properties.

Since the recycled concrete aggregate does not meet the requirements of specific gravity and content of absorbed water in JIS, it cannot be used in same as crushed stone from natural rock. However, since the recycled concrete aggregate is stronger than artificial light weight aggregate or light weight aggregate from volcanos, there is a good possibility that the recycled concrete aggregate can be used for production of concrete.

Reference

Committee of disposal and reuse of wastes from construction works, Building Contractors Society, Japan, (1975) A report for reuse of concrete demolition waste.

MANUFACTURING OF RECOVERED AGGREGATE THROUGH DISPOSAL AND RECOVERY OF DEMOLISHED CONCRETE STRUCTURES

MASAYOSHI KAKIZAKI, MINORU HARADA Kajima Corporation, Kajima Institute of Construction Technology

HISASHI MOTOYASU Fujita Construction Company, Institute of Construction Technology

ABSTRACT

This research examines effective crushing methods using a combination of various existing crushers for reinforced concrete structure waste from the demolition site of certain building erected several decades ago and have evaluated the recovered aggregate to see if it is effective as an aggregate for concrete from a grain size and quality standpoint.

Crushers used in the experiment included a jaw crusher for rough crushing, a hammer crusher for medium crushing, a roll crusher and an autogenous mill for crushing.

After crushing, recovered aggregate was found to be the same grain size as crushed stone aggregate after crushing with a jaw crusher. Furthermore, the grain size, grain distribution, specific gravity and coefficient of water absorption were improved by secondary crushing with a hammer crusher and autogenous mill, etc. This was due to the separation of cement paste attached to the surface of reused aggregate.

The quality of recovered/reused aggregate was found to be similar to the quality standard for crushed stone aggregate per JIS A 5005 "Crushed Stone for Concrete" and the quality of ordinary aggregate prescribed by "Reinforced Concrete Work JASS5" of the Architectural Institute of Japan.

Key words: Waste concrete, New recycle, Resources, Crashing machine, Recovered aggregate

1. INTRODUCTION

Waste concrete produced by demolishing concrete structures has been disposed of by burying it in reclaimed lands. Recently, however, the locations, capacity and width of reclaimed land which receives waste concrete have been limited, and has become an obstacle to demolishing work.

Also, the whole industry world has the goal of producing things without causing harm to the public and producing any waste substances. Therefore, the recovery of waste concrete, etc., which is produced in a large quantity, is one measure for resolving this problem. In addition, if this measure is used, natural resources can be saved and used highly effectively.

We studied the method for reusing waste concrete as aggregate for concrete members by crushing it into grains of proper sizes.

In this case, we crushed waste substances produced by demolishing a reinforced concrete building built several decades ago (hereinafter referred to as waste concrete) with various crushing machines (manufactured by three companies) to determine an effective crushing method, then we evaluated the grain size and quality of the aggregate produced by that method.

This study was alloted to us by the Building Constractors Society, Disposal and Recovery of Demolished Concrete Structures Committee in which we had been paticipating.

2. FEATURES AND OPERATING CONDITIONS OF VARIOUS CRUSHING MACHINES

2.1 Features of crushing machines

Since special crushers had not been developed for disposing the waste concrete, we inspected and analyzed the features of crushing machines available for solid materials from an aspect of the size of waste material input and the size of grains obtained. By considering the results, we selected proper crushing machines.

The types, model names, capacity, etc. of crushing machines used for this study are shown in Table 1.

The crushing mechanism is classified into compression, impact, shearing, bending, friction, etc. Some crushing machines use only one crushing mechanism while others use more mechanisms combined with each other. By considering the manufacture and size of aggregate, we selected the following crushing machines.

a. Crushing machine for rough crushing:

Used to crush large 300 ∿ 400 mm waste concrete blocks primarily with a 4 ∿ 8 crushing ratio*. A jaw crusher (which crushed concrete with compressive force) was used this time.

(* "Crushing ratio" is widely defined meaing as the ratio of the size of supplied material (waste concrete) to that of crushed material (recovered aggregate).)

b. Crushing machine for medium crushing:

Used to crush 50 ∿ 300 mm medium waste concrete blocks primarily and secondarily with a 4 ∿ 10 crushing ratio. Vertical and horizontal hammer crushers (which crush material with impact and compressive force) and a double-roll crusher (which crush material with both compressive and shearing force) were used.

c. Crushing machine for fine material crushing:

An autogenesis mill (which perform crushing with impact and friction) was used at this time to produce recovered fine aggregate.

2.2 Operating conditions of crushing machines

Waste concrete to be crushed consists of fine and coarse aggregate and cement paste which sticks to the former. Therefore, the characteristics of recovered aggregate depend on how much cement paste has been removed from the aggregate, which depend on which crushing method was used.

The operating conditions of each crushing machine were set properly by considering the size of waste concrete and the quality aggregate to be recovered. The operating conditions for crushing machines used for each type of recovered aggregate are shown in Table 2.

The trial manufacture plant for recovered aggregate was a crushing circuit which consisted of crushing machines, a supplier, charger and discharger of waste concrete and an aggregate separator and classifier, etc.

Table 1 Crushing machine types and main specifications

Classification of crushing machines	Employed model			Specifications for drive power	Dimensions(mm) (set)	Crushing Capacity (t/H)
	No.	Type	Manufacturer			
<For rough crushing> Jaw crusher	I	Single dogged type (OPR1530)	A	37kWx1unit, 275 rpm	380x760, O.S = 33mm	25 ∿ 40
	II	Single dogged type (3624ST)	B	75kWx1unit, 250 rpm	924x600 O.S = 60,80, 120mm	30 ∿ 60
<For medium crushing> Vertical hammer crusher	III	Composite type (KE-400)	A	150kWx2unit, v=20 ∿ 27m/s	1800x150 C.S = 55mm	40 ∿ 60
	IV	Composite type (KE-100B)	A	75kWx1unit, v=20m/s	1330x1240, C.S = 20mm	30 ∿ 40
<For medium crushing> Horizontal hammer crusher	V	Shredder type (CH6/900)	B	150kWx1unit, v=40m/s	900x900 G.B=75,150mm 3∿5 pieces	4 ∿ 10
<For medium crushing> Roll crusher	VI	Double-roll type (2430)	B	55kWx2units, 153 rpm	O.S = 25mm	Approx. 2
<For fine crushing> Autogenesis mill	VII	Single-drum type (1830 ϕ 305ℓ)	C	30kWx1unit, 26 rpm	150x150 G.B=20,30mm 6 pieces each	Approx. 0.5

v: Rotor peripheral velocity
O.S: Gap of outlet openings
C.S: Gap of choke ring
G.B: Width of grade bar

* The crushing ratio is generally defined in a wide meaning as the ratio of the size of supplied material (waste concrete) to that of crushed material (recovered aggregate.)

Table 2 Types of recovered aggregates produced under various conditions

Type No. of re-covered aggre-gate	Machines and conditions for primary crushing		Machines and conditions for secondary crushing	
	Crushing machine	Operating condition	Crushing machine	Operating condition
1	I -Jaw crusher	O.S : 33	-	-
2	II -Jaw crusher	O.S : 60	-	-
3	II -Jaw crusher	O.S : 80	-	-
4	II -Jaw crusher	O.S :120	-	-
5	III -Composite type KE-400	v : 20	-	-
6	III -Composite type KE-400	v : 27	-	-
7	IV -Composite type KE-100B	v : 22	-	-
8	V -Shredder CH6/900	v : 40	-	-
9	II -Jaw crusher	O.S : 60	V -Shredder CH6/900	v : 40
10	II -Jaw crusher	O.S : 80	V -Shredder CH6/900	v : 40
11	II -Jaw crusher	O.S :120	V -Shredder CH6/900	v : 40
12	II -Jaw crusher	O.S :120	VI-Roll crusher	O.S : 25
13	VII -Autogenesis mill	Batch type, 20 min	-	-
14	VII -Autogenesis mill	Continuous type. G.B:30	-	-
15	VII -Autogenesis mill	" G.B:20	-	-

O.S : Gap of outlet opening (mm), v : Rotor peripheral velocity (m/s), G.B : Width of grade bar

3. PROPERTIES OF RECOVERED AGGREGATE

Normal aggregate (natural aggregate and crushed stone), natural lightweight aggregate and artificial aggregate are generally used for concrete. Crushed stone is produced by properly crushing round stones or rocks, and the production line for this aggregate is similar to that for making aggregate from waste concrete.

The necessary quality of aggregate for concrete including the grain size, specific gravity, water absorption, solid volume percentage for grain size judgement, etc. must be kept constant in the production line.

The criteria for aggregate quality is changed little according to the type, scale, strength, etc. of a concrete structure for which aggregate is used.

The quality of recovered aggregate was inspected according to JIS A 5005 "Crushed stone for concrete" which is the quality standard for crushed stone used as aggregate and "Work on reinforced concrete structure JASS 5" established by the Architectural Institute of Japan.

3.1 Grain size distribution of recovered aggregate

The grain size distribution of recovered aggregate obtained with various crushing tests (Table 2) is shown in Fig. 1 ∿ 4 (Average of values obtained at eight testing places). JASS-5 standard gain size distribution of coarse aggregate (size under 25mm) and fine aggregate is additionally shown in the same figures.

The change in the grain size distribution of aggregate has a great effect on the workability, bleeding, and unit water quantity of concrete.

The result of evaluating each recovered aggregate based on the quality standard specified by JIS A 5005 and JASS 5 is shown in Table 3. Table 3 shows that the batch-type crushing method (No. 13) with an autogenesis mill is suitable for the production of recovered fine aggregate.

In the case of the production of recovered coarse aggregate only by primary crushing, a jaw crusher with a 60 mm (No. 2) opening, a (No. 8) and an autogenesis mill (No. 14) are suitable. Also, a combination of primary crushing with a jaw crusher with an 80 ∿ 120 mm opening and secondary crushing with a hammer crusher or roll crusher (No. 11, 12) can be used.

The large size recovered coarse aggregate than JASS 5 is sometimes able to conform to JASS 5 if it is processed again with a crushing machine (example: autogenesis mill).

3.2 General properties of recovered aggregate

The average values of specific gravity, water absorption, weight of unit volume, solid volume percentage and fineness modulus of the recovered aggregate are shown in Table 4.

The absolute density of recovered fine aggregate is 2.0 ∿ 2.7, and is 4 ∿ 20% lower than that of natural commonly used fine aggregate (river sand, land sand, sea sand, etc.: 2.5 ∿ 2.8), but recovered fine aggregate No. 9 is conforming to JASS 5.

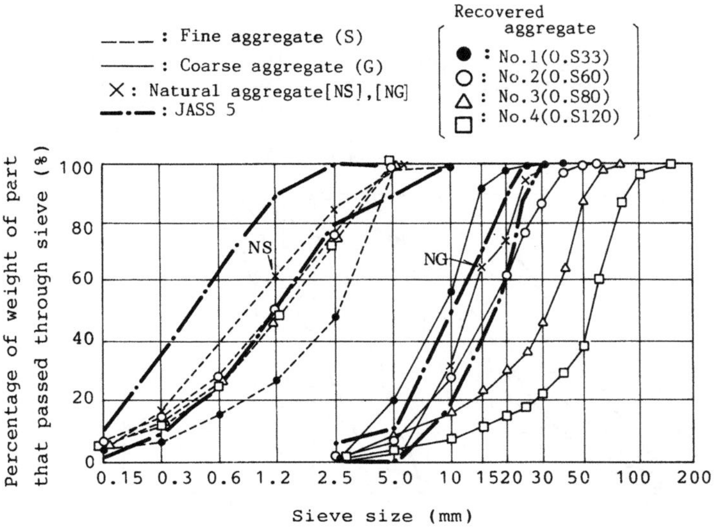

Fig.1 Grain size distribution of recovered aggregate made under various operating conditions with a jaw crusher

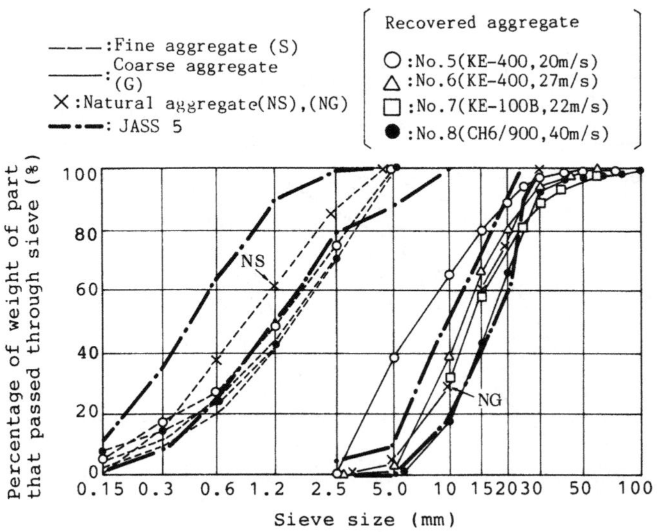

Fig.2 Grain size distribution of recovered aggregate made under various operating conditions with a hammer crusher

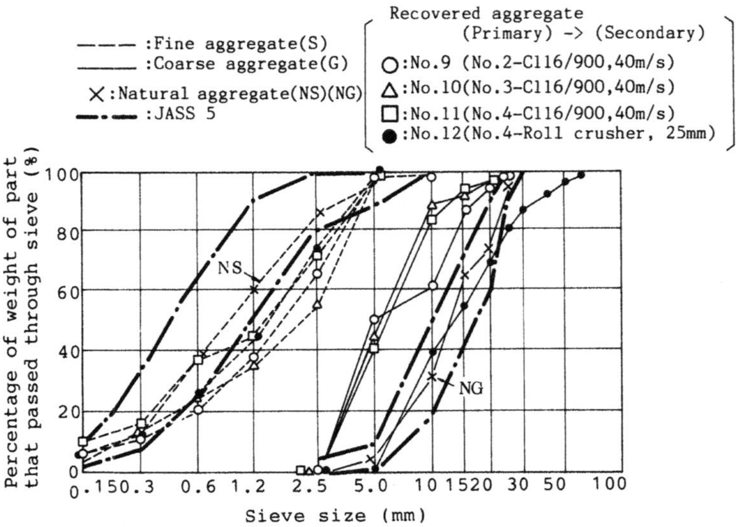

Fig.3 Grain size distribution of recovered aggregate made by secondary crushing with a hammer and roll crushers

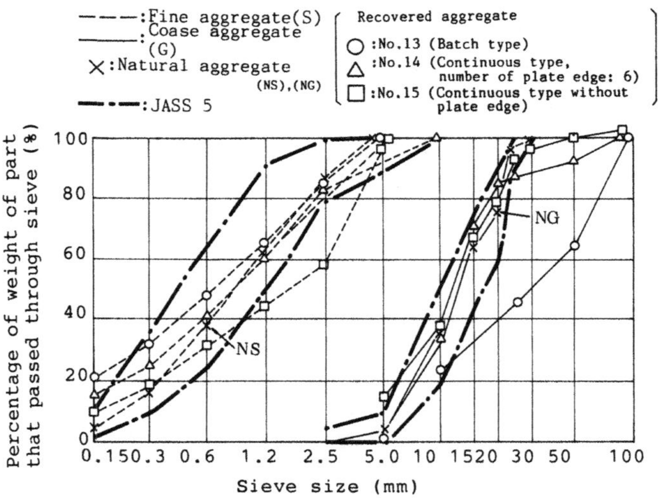

Fig.4 Grain size distribution of recovered aggregate made under various operating conditions with an autogenesis mill

Table 3 Evaluation of grain size distribution for recovered aggregate (Standard)

Standard / Aggregate Type No.	JIS A 5005 (Crushed stone)	JASS 5 Normal (aggregate)	
	Coase aggregate	Fine aggregate	Coarse aggregate
1	Crushed stone 1505,	×	×
2	Crushed stone 4005,	△	△
3	(Crushed stone 5005*, 4005*)	△	×
4	-	△	×
5	-	×	×
6	-	×	×
7	-	×	×
8	(Crushed stone 4005*)	×	○
9	-	×	×
10	(Crushed stone 2506*)	×	○
11	(Crushed stone 2505*)	×	○
12	(Crushed stone 4005*)	×	○
13	-	△	×
14	-	○	○
15	Crushed stone 2505	×	○

-: Does not confirm with any standard.
(*): Similar products
○ : Conforming to JASS 5
△ : Nearly conforming to JASS 5
× : Unconforming to JASS 5

On the other hand, the absolute density of recovered coarse aggregate is 1.9 ∿ 2.4, and is 8 ∿ 27% lower than that of natural coarse aggregate (2.6).

The water absorption of recovered fine aggregate is 9.4 ∿ 11.8%, and is 2.8 ∿ 3.6 times as high as that of natural fine aggregate (3.3%). The water absorption of recovered coarse aggregate is 3.5 ∿ 6.3%, and is 2.5 ∿ 4.5 times as high as that of natural coarse aggregate (1.4%).

The water absorption of recovered aggregate is much higher than that of natural aggregate because the water absorption of cement paste contained in the recovered aggregate is very high. The water absorption of recovered coarse aggregate does not rarely conform to JASS 5 (water absorption of coarse aggregate and fine aggregate is respectively maximam 3.0% and 3.5%). But recovered aggregate No.6 and 7 are nearly JASS 5.

The respective solid volume percentages for recovered fine and coarse aggregate are almost the same as natural aggregates. The solid volume percentages for recovered coarse aggregate No. 2, 5 and 14 conform to JASS 5005.

3.3 Application to recovered concrete

It was proved that recovered aggregate made by crushing concrete blocks with a jaw crusher had a grain diameter similar to that of crushed stone. The grain diameter, grain size distribution, specific gravity and water absorption of recovered aggregate were improved by crushing it secondarily with a hammer crusher, roll crusher, autogenesis mill, etc. This improvement apparently occurred because the cement paste which had stuck to recovered coarse aggregate was removed.

Since mortar mixed in recovered coarse aggregate greately affects the strength of recovered concrete, the production of recovered fine aggregate by using a water jet, ball mill, etc. should be examined.

Table 4 Results of tests on the general properties of recovered aggregate

Type of aggregate \ Properties of aggregate			Specific Gravity (Absolute dry)	Water absorption (%)	Weight of unit volume (kg/ℓ)	Solid volume percentage (°/vl)	Fineness modulus (%)
Fine aggregate	[NS]		2.53	3.31	1.77	67.5	2.91
	Recovered aggregate [CS]	1	2.03	10.57	1.37	62.8	3.42
		2	2.01	11.83	1.36	64.9	3.33
		3	2.00	11.70	1.33	-	3.33
		4	2.04	10.00	1.29	-	3.06
		5	2.06	10.16	1.38	64.0	2.86
		6	2.09	9.96	1.35	60.6	3.51
		7	2.14	8.30	1.55	-	3.29
		9	2.65	11.40	1.37	-	3.51
		10	2.06	10.50	1.37	-	3.67
		11	2.11	9.40	1.24	58.8	3.73
		14	2.11	10.46	1.49	68.0	-
		15	2.10	9.75	1.39	63.4	2.96
Coarse aggregate	[NG]		2.63	1.37	1.74	65.1	6.90
	Recovered aggregate [CG]	1	2.34	4.80	1.31	49.4	6.15
		2	2.27	5.96	1.41	64.8	6.54
		3	2.29	5.81	1.44	-	6.73
		4	1.89	6.37	1.37	-	7.24
		5	2.34	5.21	1.44	63.6	6.10
		6	2.42	3.93	1.39	50.4	5.84
		7	2.44	3.52	1.50	-	5.25
		9	2.26	6.15	1.42	-	6.27
		10	2.26	5.30	1.33	-	6.59
		11	2.32	5.24	1.22	52.6	6.73
		14	2.35	5.24	1.54	59.1	-
		15	2.33	5.19	1.30	52.9	6.03

[NS] : Natural fine aggregate [NG] : Natural coarse aggregate

4. NOISE FROM CRUSHING MACHINES

The noise levels of various crushing machines used are shown in Table 5. The noise levels that occurred when crushing concrete blocks with those machines were similar to those which occurred during common demolishing of concrete structures (At 30 m, CCR (low-exploding speed crushing powder): 83 phon, Large-sized breaker: 90 phon).

When recovering aggregate from concrete blocks in a city area or around a residential area, the noise level of crushing machines must be lowered below those regulated by the Noise Control Regulation or noises which must be arrested by a noise prevention fence, noise prevention housing, etc.

Table 5 Noise levels of various crushing machines during concrete crushing

Crushing machine (Type)	Capacity of drive power	Operating condition	[SPL]2m(Phon) 2)		[SPL] 3) 30m (Phon)	[PWL] 4) (dB)
			No load	When crushing concrete		
Jaw crusher 1) (Single dogged type, Model HA-5008)	45kW x 1 unit	Gap of outlet opening: 25∿40mm	65∿72	82∿85	56∿62	99
Jaw crusher (Single dogged type, Model 3624 ST	75kW x 1 unit	Gap of outlet opening: 60∿120mm	80∿84	92∿93	69∿70	107
Vertical hammer crusher (KE-100B)	150kW x 2 unit	Rotor peri-pheral velocity: 20 m/s	90 ∿102	100∿115	77∿92	129
		Rotor peri-pheral velocity: 27 m/s	90 ∿98	100∿116	77∿93	130
Vertical hammer crusher (Com-posite type, KE-400)	74kW x 1 unit	Rotor peri-pheral velocity: 22 m/s	85 ∿99	95∿110	72∿87	124
Horizontal hammer crusher (Shredder type, CH 6/900)	150kW x 1 unit	Rotor peri-pheral velocity: 40 m/s	85∿88	First 100∿102	77∿79	116
		Rotor peri-pheral velocity: 40 m/s	85∿88	Second 94∿98	71∿75	112

1) Noise of jaw crusher manufactured by D company
2) $[SPL]_{2m}$: Noise level at 2m from machine body
3) $[SPL)_{30m}$: Noise level at 30m from machine body
4) [PWL] : Acoustic power level of crushing machine

5. CONCLUSION

Since the reuse of waste concrete (estimated quantity: 3,000,000 ∿ 4,000,000 m^3/year) is a new recycle industry for resources, studies in this field must be further developed for applying the crushing method just explained in practice.

To accomplish this, a special crushing machine for waste reinforced concrete and a production system for recovered aggregate must be developed, and studies on the prevention of noise and vibration at production plants and secondary air polution produced by dust must be further promoted.

DEMOLITION WASTE OF THE "ZANDVLIET" LOCK AS AGGREGATES FOR CONCRETE.

D. MORLION
M.B.G. - C.F.E.,Belgium
J. VENSTERMANS and J. Vyncke
Belgian Building Research Institute (C.S.T.C. - W.T.C.B.), Brussels

Abstract
For the construction of part of the embankment walls of the "Berendrecht" lock, the demolition waste of the "Zandvliet" lock was used. In function of the properties of the produced rubble aggregates an ideal composition for the recycled concrete had to be determined. During the processing of the demolition waste due attention was paid to the grading, the filler content, the particle size index and the water absorption of the produced rubble aggregates. The mix proportions of the recycled concrete were determined on the basis of an extensive study. A concrete grade B35 was perfectly well attainable with the rubble aggregates. Further the shrinkage and workability of the recycled concrete was found to be nearly similar to this of the conventional concrete which was used at other locations of the site.
Key words: Recycled concrete, Rubble Aggregates, Workability, Compressive strength, Shrinkage, Mix proportions.

1. INTRODUCTION

In view of the expansion of the port of Antwerp it was decided to build nearby the old "Zandvliet" lock a new lock ("Berendrecht") with greater capacity. For the construction of this worldwide largest lock 650,000 m^3 of concrete had to be cast. As a common access channel to the "Zandvliet" and "Berendrecht" lock was foreseen the old embankment walls of the "Zandvliet" lock had to be demolished. The demolition of this walls was done by means of explosives and yielded 80,000 m^3 of demolition waste. As a result of economical and ecological considerations it was decided to look for possibilities of using the waste materials for recycled concrete. With this aim a study project, under the guidance of the Department for the Development of the left bank of the river Scheldt, was commissioned by the Belgian Ministry of Public Works to the Belgian Building Research Institute and the joint venture "Berendrechtsluis".

2. DESCRIPTION OF THE AVAILABLE CONCRETE RUBBLE

Three different concrete grades can be distinguished in the available rubble i.e.:

- concrete containing gravel 0/32,
- concrete containing gravel 0/64,
- and the concrete of the cast in place foundation piles.

Visual inspection of the rubble reveals that part of it, especially the rubble derived from the foundation piles, is contaminated with loam and clay.

In order to characterise the quality of the concrete rubble the compressive strength of the concrete was determined on small cube-samples (100x100x100mm^3), extracted from the demolition waste. Additionally, for comparative purposes, impact hammer testing (Schmidt rebound hammer) was performed at various locations in the stock.
The characteristic compressive strength of the concrete was found to be equal to approximately 30 N/mm^2, while the standard deviation is about 6 N/mm^2.

Seemingly, for mass concrete these waste materials can thus very suitably be used.

3. QUALITY CONTROL OF THE RUBBLE AGGREGATES

During the installation an optimisation of the rubble processing plant due attention was paid to the following aspects:
- the compliance of the gradation of the main part of the produced rubble aggregates with the specified requirements,
- the presence of contaminants,
- the particle shape index,
- the water absorption of the rubble aggregates.

3.1. Grading

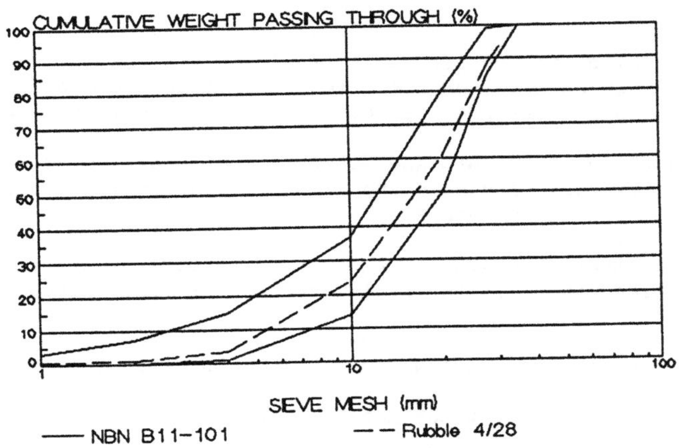

Fig. 1: Sieve analysis of the coarse aggregates.

The production of recycled coarse aggregates 4/28 was the major objective. Several sieve analysis of samples taken at the crusher revealed that the particle size distribution of the obtained coarse aggregates was in accordance with the specifications of the Belgian standard NBN B11-101 (Fig. 1).
In parallel to the mentioned production two other aggregate fractions where obtained i.e. fine aggregates 2/4 and recycled sand 0/2. However, due to the presence of to much cement particles, these fractions did not comply with the requirements specified in the former mentioned standard (Fig. 2).

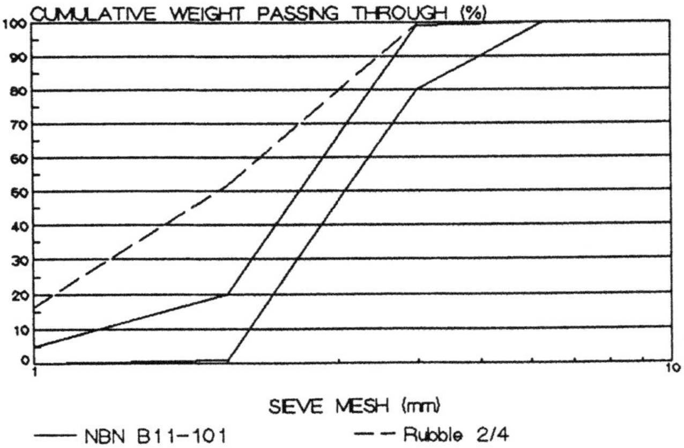

Fig. 2: Sieve analysis of the fine aggregates.

3.2. Contaminants

The presence of contaminants, particularly finely pulverized material (filler) and clay, can have an adverse effect upon concrete quality. These substances must therefore as effectively as possible be removed.
The demolition waste had to be dredged from the bottom of the river Scheldt. Inevitability very fine sand and silt which are mainly found over there, got mixed with the rubble.

On the basis of comparative studies and an extensive literature survey it was decided to limit the filler content of the produced aggregates to 0.3%.

The filler content of the material was determined in the following way:
- A representative test sample of the material is washed through the sieves with mesh size 2mm and 74 μ.
- The material retained on the 2mm sieve is then oven-dried and weighed (P_1).

- Also the material passing the 2mm sieve but retained on the sieve with mesh size 74µ is oven-dried and weighed (P_2).
- The water containing the material finer than 74µ is allowed to settle for a few hours. After decanting the clear water the oven-dry weight (P_3) of the residue is determined.
- As some fines may have adhered to the larger aggregates the oven-dried material retained on the 2mm sieve is now sieved dry. The material finer than 74µ is denoted P'_1, the coarser part passing the 2mm sieve but retained on the 74µ sieve is denoted P''_1.
- Defining $P_5 = P'_1 + P_3$ and $P_4 = P_4 + P''_1$ the amount of filler is calculated as $\frac{P_5}{P_1+P_2+P_3}$. The value of $\frac{P_4+P_5}{P_1+P_2+P_3}$ is called the sand content of the aggregates.

For the first samples following values were found:
- filler content = 1.03%,
- sand content = 4.40%.

In order to reduce the filler content water sprinklers were mounted over the belt conveyors and sieves. Fairly good results were obtained in this way as can be seen in table 1. If necessary the material could be passed several times over the sieves or the length of the loop over the conveyors could be changed.

Table 1: Determination of the filler content of rubble aggregates.

Sample No	P_1 (g)	P_2 (g)	P_3 (g)	P_4 (g)	P_5 (g)	Sand content (%)	Filler content (%)
1	14302.7	262.6	30.2	302.5	42.7	2.37	0.29

For practical reasons, in the field, the approximate amount of filler was determined as $\frac{P_3}{P_1+P_2+P_3}$.

So, it was possible within 24 hours to have the results concerning the filler content of the produced material. Tests for the filler content were made systematic during the whole production process.

3.3. Particle shape index

The presence of flat and elongated particles in the coarse aggregates can be expressed by the particle shape index. The particle shape index for the rubble aggregates has been verified and was found to be nearly the same as for natural coarse aggregates.

3.4. Water absorption

Rubble aggregate particles can absorb relatively large amounts of water. This must be taken into account during the processing of the concrete mix, more particularly from the viewpoint of ensuring its workability for a sufficient length of time. Several tests were performed to determine the water absorption capacity of the aggregates. Also the time it takes to saturate the aggregates by immersion in water was determined (Table 2 & Fig. 3).

Table 2: Saturation and water absorption of the aggregates.

Sample	Water absorption (% of dry mass)	Time to saturation (minutes)
coarse aggregates 4/28	5	15
fine aggregates 2/4	10	10
sand 0/2	17	5

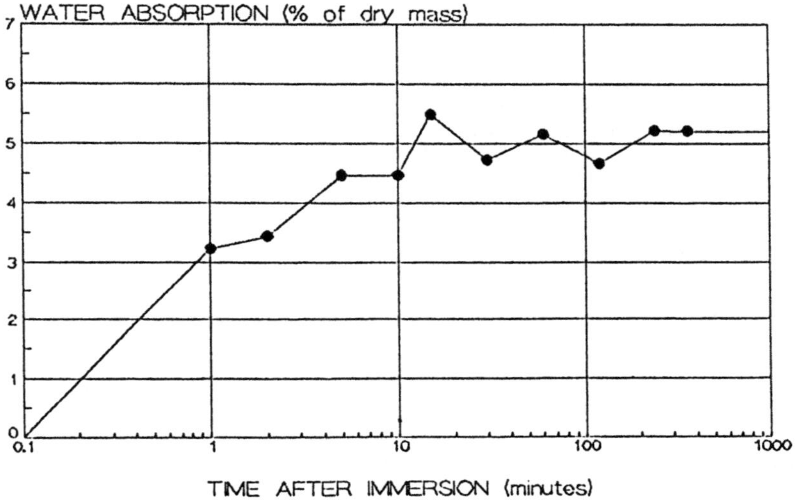

Fig. 3: Water absorption in function of time - aggregates 4/28.

4. CONCRETE MADE WITH RUBBLE AGGREGATES

4.1. General observations

The recycled concrete was to be used as mass concrete for the construction of part of the embankment walls and footings of the access channel of the new lock and had to be equivalent to the conventional concrete made with gravel used at other locations of the site. Especially the workability, the strength and the deferred deformations of the reference concrete and the recycled concrete had to be nearly equal.

In order to reduce the shrinkage the water/cement ratio had to be kept as low as possible. A cement content of 350 kg/m^3 was further imposed by the requirements of the project specifications.

To avoid colour differences between the parts constructed with recycled concrete and conventional concrete the same cement had to be used for both types of concrete. To control the temperature rise in the mass concrete due to the heat of hydration the use of a low-heat type portland blastfurnace cement LK30 was imposed.

The concrete mix proportions for the conventional concrete were as follows:

- Gravel 3/28 1063kg
- Sand 0/5 (Rijn) 378kg
- Sand 0/2 (Schelde) 285kg
- Cement LK30 350kg

Making use of the rubble aggregates available on the plant the following types of recycled concrete could be produced:

- Type 0, where only part of the conventional coarse gravel is replaced by recycled coarse rubble aggregates 4/28,
- Type I, where the overall quantity of coarse gravel is replaced by recycled coarse aggregates 4/28,
- Type II, where as well the coarse gravel as the fine gravel are replaced by recycled aggregates 4/28 and 2/4,
- Type III, where all aggregates are replaced by recycled materials 4/28 2/4 and 0/2.

However, as the rubble aggregates are of rather good quality the first possibility was not envisaged.
An extensive test program was set up to determine the ideal concrete composition, the strength and the deformation characteristics of the recycled concrete. With reference to economic aspects the composition which fulfilled the requirements of the project specifications at the lowest cost per m^3 of fresh concrete was obvious the best choice.

4.2. Theoretical determination of ideal concrete compositions

For the concrete types I,II and III theoretical ideal concrete compositions were determined using two different approaches: i.e. the method of "Dreux-Gorisse" and the method of "Fuller". Whereas the latter method only depends on the maximum grain size the former takes account of among others the shape of the particles, the effectiveness of the consolidation equipment, the cement content and so on...
According to the performed calculations the mix proportions given in table 3 could be used.

Table 3: Theoretical ideal concrete mixes.

	Mix proportions (kg/m^3)					
Approach	DREUX-GORISSE			FULLER		
Concrete type	I	II	III	I	II	III
Cement LK30	350	350	350	350	350	350
Natural sand 0/2 (Schelde)	123	62	-	19	141	125
Recycled sand 0/2	-	-	287	-	-	215
Natural sand 0/5 (Rijn)	461	235	-	611	178	-
fine rubble 2/4	-	300	336	-	407	382
coarse rubble 4/8	1215	1189	1103	1098	985	985

As can be seen, according to the method of Fuller, it appears impossible to use only recycled aggregates.

4.3. Practical determination of the concrete composition

As the two theoretical approaches yield rather inconsistent results and experimental test program was set up. Various concrete mixes were prepared and compared one to another on basis of their compressive strength.

For the concrete type I, apart from the two theoretical compositions I-F (Fuller) and I-D (Dreux-Gorisse) six other concrete mixes (I-3 up to I-8) were tested (table 4). For the concrete types II and III five respectively three concrete mixes were tested (table 5 & 6).

Table 4: Concrete mixes type I concrete.

Batch	Mix proportions (kg/m^3)			
	Cement LK30	Natural sand 0/2 (Schelde)	Natural sand 0/5 (Rijn)	Coarse rubble aggregates 4/28
I-F	350	40	600	1116
I-D	350	120	450	1187
I-3	350	60	500	1200
I-4	350	100	460	1200
I-5	350	40	570	1150
I-6	350	120	540	1100
I-7	350	140	620	1000
I-8	350	50	710	1000

Table 5: Concrete mixes type II concrete.

Batch	Mix proportions (kg/m^3)				
	Cement LK30	Natural sand 0/2(Schelde)	Natural sand 0/5 (Rijn)	Fine rubble aggregates 2/4	Coarse rubble aggregates 4/28
II-F	350	145	183	418	1011
II-D	350	61	231	295	1170
II-3	350	86	319	194	1158
II-4	350	-	364	383	1011
II-5	350	511	-	234	1013

Table 6: Concrete mixes type III concrete.

Batch	Mix proportions (kg/m^3)				
	Cement LK30	Natural sand 0/2(Schelde)	Recycled sand 0/2	Fine rubble aggregates 2/4	Coarse rubble aggregates 4/28
III-F	350	129	221	393	1014
III-D	350	-	292	342	1123
III-3	350	-	266	382	1109

For all mixes 350kg cement per m^3 concrete was used. The water cement ratio for the theoretical mixes was taken equal to 0.55 while for the

other mixes it was adjusted in such a way that for all batches the same slump was attained. Batching was done manually and due attention was taken of the moisture content of the aggregates.
It is to be noted that for practical reasons the amount of natural sand 0/2 lightly has been changed for the composition I-F.

4.4. Compressive strength of the recycled concrete

The compressive strength of the various types and compositions of recycled concrete was determined at 7 and 28 days.
Table 7 gives the results of the compression tests.

Table 7: Compression tests - Test results.

Batch	I unit weight (kg/m^3)	I compressive strength (N/mm^2) 7d	I compressive strength (N/mm^2) 28d	II unit weight (kg/m^3)	II compressive strength (N/mm^2) 7d	II compressive strength (N/mm^2) 28d	III unit weight (kg/m^3)	III compressive strength (N/mm^2) 7d	III compressive strength (N/mm^2) 28d
F	2294	34.8	39.5	2240	27.0	31.6	2174	17.7	22.8
D	2317	31.4	32.9	2173	21.6	26.0	2174	22.1	25.0
3	2300	31.1	32.4	2284	28.0	32.9	2173	16.7	22.4
4	2299	31.7	31.9	2243	25.3	29.3			
5	2342	30.3	32.7	2259	25.9	28.9			
6	2300	34.1	35.9						
7	2302	30.8	32.9						
8	2287	33.9	32.9						

4.5. Shrinkage

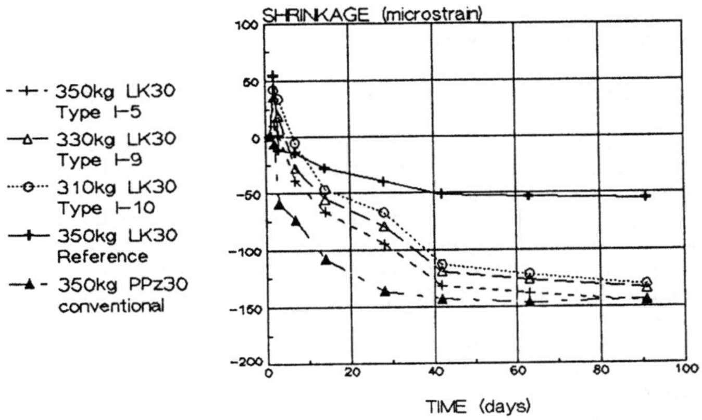

Fig. 4: Shrinkage in function of time.

As it is well known that the shrinkage of concrete made with rubble aggregates may be greater than that for comparable conventional mixes some exploratory tests were performed.
Comparisons were made between the shrinkage of:
- the conventional reference concrete with 350kg LK30 cement
- a conventional concrete analogue to the reference concrete with Portland-Pozzolan cement PPz30 instead of LK30 cement
- the recycled concrete I-5 with 350kg LK30 cement
- a recycled concrete analogue to I-5 with 330kg LK30 cement instead of 350kg LK30 cement (I-9)
- a recycled concrete analogue to I-5 with 310kg LK30 cement (I-10)

The shrinkage test were performed in accordance with the Belgian standard NBN B11-216. Fig. 4 gives the obtained results.

4.6. Conclusions

Considering the compressive strength of the various concrete mixes it may be concluded that the mixes I-F and I-6 yield very good results. A concrete grade B35 is indeed perfectly well attainable with these mixes, which by the way have nearly the same coarse aggregate content. Low compressive strengths are obtained with the mixes II and III. This was to be expected in view of the rather large filler content of the fine rubble aggregates and the recycled sand. It is to be remembered however that the main objective of the rubble processing plant was to produce coarse aggregates 4/28.
As expected it is seen that in comparison to the reference concrete somewhat larger values are found for the shrinkage of the rubble concrete. However this poses for the actual case no problems. The shrinkage is indeed limited to 150µm/m and is even smaller than for the conventional PPz30 concrete.
For the construction of the embankments it was decided, on the basis of the test results, to use the concrete mix I-F.

5. RECOMMENDATIONS AND GENERAL CONCLUSIONS

The following recommendations should be kept in mind regarding recycled concrete:
- Special care has to be taken as to the presence of contaminants in the recycled aggregates, not only the presence of filler but also the sand content must be limited.
- Rubble aggregate particles can absorb relative large amounts of water. This must, as already mentioned, be taken into account during the processing of the concrete mix. To prevent a rapid decrease in workability rubble aggregates must be sufficient wet before concrete mixing takes place. Immersion of the aggregates in water during about one hour prior to mixing turned out to be a very efficient solution.
- The amount of water to be added to the batch must be corrected in function of the consistency measurements. Therefore the consistency of the fresh concrete must be monitored for example on the Watt meters of the mixers.

Taking into account these latter recommendations there was found to be little difference in the workability of the recycled concrete and the conventional concrete.

References

Morlion, D. and Balcaen, M. (1987) Kringloopbeton afkomstig van de afbraak van oude kaaimuren voor de bouw van de nieuwe kaaimuren. Studiedag KVIV & WTCB-CSTC, Antwerpen, pp 1-37.

THE USE IN ROADS OF AGGREGATES MADE FROM DEMOLITION MATERIALS

M. BAUCHARD
Laboratoire Régional de l'Est Parisien, Le Bourget - France

Abstract

The recycling of demolition materials for use in roads has been developed in two parts of France where "deposits" of such materials are especially abundant : the Paris area and the Nord-Pas-de-Calais region.

The first tests were conducted in 1976 on the A1 motorway near Paris. Between 1981 and 1986, seven permanent crushing-screening plants (five in the Paris area, two near Lille) were set up, generally at locations close to mixing plants, to prepare road-building materials treated with hydraulic binders. The total output of crushed-concrete aggregates is now about 600,000 tons a year. Thanks to the work of the equipment builders and careful thinking about the design and adjustment of the crushing-screening plants, the quality of the aggregates has been significantly improved in a few years and materials that can be used in the subbase and roadbases of pavements designed for moderate traffic levels are now available.

Roadbases for pavements designed for heavy traffic have also been built, on an experimental basis, to evaluate the mechanical strength limits of these special aggregates.
<u>Key words</u> : Aggregates Recycling, Concrete, Road Construction, Road Base, Crusher, Pavement.

1. Historical background

After the Second World War, a number of European countries, notably Germany, England and the Netherlands, made systematic attempts to re-use demolition materials in the reconstruction of civil engineering works. Articles published at the time describe these attempts and cite the first results of investigations of aggregates made of crushed concrete. After this postwar period, there was a rather long pause in these studies, broken since 1973 by the publication of many articles by American authors (technical, economical, and feasibility studies, descriptions of projects).

The first experimental work in France was done in 1976 on the A1 motorway in the Paris area, where aggregates produced by crushing the concrete of the old pavement were re-used in a lean concrete constituting the subbase of the new pavement. This approach is now used routinely in the Paris area.

Following these various experiments, a large step remained to be taken: designing fixed plants capable of producing aggregates from a great variety of demolition materials. The development of such equipment was begun in 1980 at the La Villette site in Paris (demolition of slaughterhouses) and continued until 1985, when a fixed plant was set up by the Societe D.L.B. at Limeil-Brévannes in the Val-de-Marne department.

In the meantime, other construction contractors and a number of French equipment builders were designing other installations. The Paris area is a good place to develop the use of recycled materials as aggregates, primarily because the distance to the nearest sources of natural materials is increasing, and now has six recycling plants. The Lille area, another large conurbation, also has a large recycling plant, at Fretin.

2. General characteristics of the installations in the Paris area

Without going into technical details, we may distinguish two types of installation :

2.1 Those that produce aggregates by primary crushing

These are the simplest installations : the aggregates are produced by a single crushing operation. As a result, their geotechnical properties depend to a large extent on the quality of the demolition materials put into the crusher.

From the four installations of this type now in service, we find that the demolition materials are in fact carefully selected (reinforced or unreinforced concrete, for example).

This basic precaution ensures that the quality of the aggregates will be adequate for the planned utilizations.

The mode of preparation of the aggregates is the same (impact crushing) at these four installations, but the equipment is different:

- Bergeaud crusher at the Société MATRIF (Vigneux)
- Blaw-Knox crusher at the Société SLAM (Sucy-en-Brie)
- Hazemag crushers at the Société YVES PRIGENT (Emerainville) and the Société S.D.V.M. (Valenton).

These installations also differ in their design details. In addition, modifications or additions aimed at improving output quality are common.

2.2 Those that produce aggregates by primary and secondary crushing

Two such installations are now in service :

- At the Société D.L.B. at Limeil-Brévannes (basically Hazemag equipment)
- At the Société S.N.M. at Gennevilliers (Neyrtec equipment).

These are more complex fixed plants designed for the processing of demolition materials having varied origins. It should however be noted that D.L.B. does not currently make use of this possibility: it crushes only reinforced or unreinforced concrete.

In their overall design, these installations are not very different from, say, Dutch installations. But the French engineering departments involved (equipment builders and contractors) have designed specific items of equipment that are exactly matched with local needs and meet the quality standards imposed for various road-building applications.

3. General properties and cost of the aggregates

The materials consist of a mixture of the original aggregates and grains of crushed mortar. As a result, their hardness properties are depending on the origin of the demolition materials (buildings or pavements), the Los Angeles coefficient ranges from 29 to 35 and the Micro-Deval from 15 to 24. The true density of these materials is naturally lower than that of the usual materials, because of the high porosity of the particles of mortar.

Automatic devices eliminate certain undesirable materials (magnetic sorters for reinforcements; "Aquamator" type processes for low-density residues), completing the work done at a manual sorting station.

In practice, it is found that these operations are being mastered better and better, but are still the weak point of the recycling installations.

In addition to the various technical constraints on the preparation of these materials, costs of personnel are high, equipment wear is high, and shutdowns of the installations are common. The costs of preparing these materials may be twice as high as the costs of preparing silico-calcareous materials: roughly, between 40 and 50 francs a ton.

In addition to the saving achieved through the use of these aggregates, there is another saving arising from the low density of the products prepared by the mixing plants.

The figures below, which should be regarded as orders of magnitude, show the importance of the "tonnage in place" parameter.

	Silico-calcareous formula (tons/m³)	Crushed-concrete formula (tons/m³)	Saving on tonnage in place
Untreated aggregate	2.30	2.10	8.7%
Aggregate-cement	2.35	2.16	8 %

4. Geotechnical properties of the materials from the various installations

The current French standard applies "to aggregates of natural or artificial origin used in the concrete industry, in building and public works..." The aggregates produced from demolition materials therefore fall within the scope of this standard, which in fact makes it easier to classify them.

4.1 Materials prepared by the D.L.B. plant (Limeil-Brévannes)
This is the oldest of the plants. It bears the full burden of the development work. Since its commissioning, the secondary crusher has been replaced twice and the primary crusher once.

The geotechnical properties stated below are for recent output (two months of monitoring) of 0/6mm and 6/20mm aggregates produced by the secondary crushing of reinforced and unreinforced concretes.

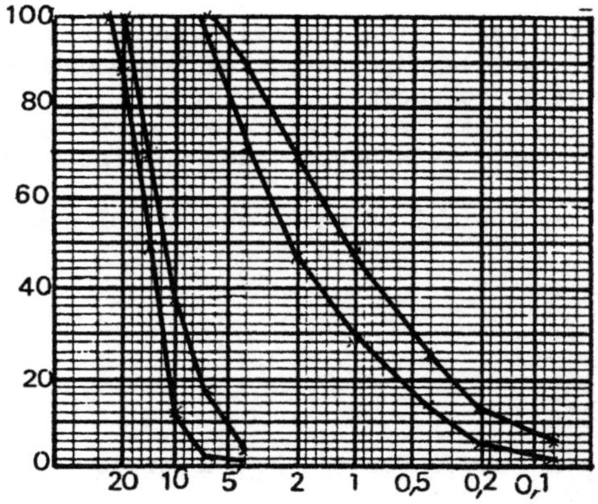

<u>0/6mm sand</u>
.14 < Friability number < 17
.54 < Sand equivalent < 68
.Proposed classification : b

<u>6/20mm gravel</u>
.29 < Los Angeles < 34
.15 < Micro-Deval < 24
.Proposed classification: D II*

*This classification takes account of the rule for a trade-off between the Los Angeles and Micro-Deval coefficients given in standard NF P 18-321

In addition, the arrangements made as regards the collection of the demolition materials make it possible to produce an unreconstituted 0/30 - 0/40mm material of highly satisfactory quality by primary crushing.

4.2 Materials prepared by the S.N.M. plant (Gennevilliers)
This installation was commissioned recently.

In its final stage of design and adjustment, this plant took advantage, to some degree, of the experience gained at the D.L.B. plant, although its technical and commercial objectives were substantially different.

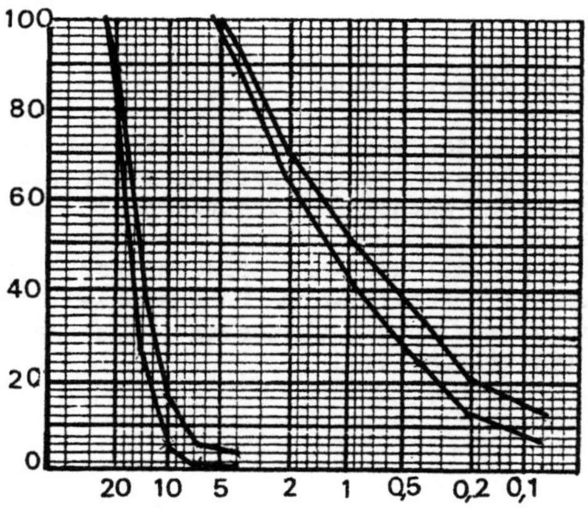

<u>0/6mm sand</u>
.15 < Friability number < 18
.28 < Sand equivalent < 61
.Proposed classification : b

<u>6/20mm gravel</u>
.29 < Los Angeles < 35
.19 < Micro-Deval < 23
.Proposed classification: D II*

*This classification takes account of the rule for a trade-off between the Los Angeles and Micro-Deval coefficients given in standard NF P 18-321

Priority has been given to the production of unreconstituted screened 0/30mm materials from pavement demolition materials (aggregate-cement, aggregate-slag, and foundation courses).

The balance of the output consists primarily of 40% 0/6mm and 6/20mm secondary-crushing aggregates that seem to have geotechnical properties similar to those of the D.L.B. aggregates.

4.3 Materials prepared by the Yves Prigent plant (Emerainville)

The materials offered by this plant are primary crushing products screened on different sieves according to needs.

The quality and uniformity of these materials are ensured by a rigorously selective collection of reinforced and unreinforced concrete.

The characteristics of one of the typical products are as follows :

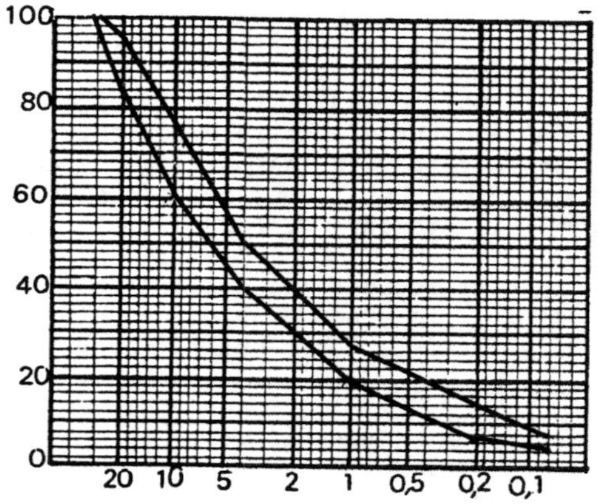

0/31.5mm crushed-concrete aggregate

46 < Sand equivalent < 64

Construction contractors think very highly of this material, which is not very different from those produced by primary crushing in the D.L.B. and S.N.M. plants, because it stands up well to site traffic and is relatively unaffected by water.

5. The use of crushed-concrete aggregates in roads

There are a number of documents to which project supervisors, engineering departements, and professional organizations refer that propose specifications for pavement materials according to the importance of the structures to be built.

In France, the reference document used by the network of Laboratoires des Ponts et Chaussées (Public Works Laboratories) is "Spécifications relatives aux granulats pour chaussées" (Pavement aggregate specifications - April 1984 directive).

Strict application of this document gives the following table :

Type of pavement		Traffic class * > T2	T2	T3	< T3
Pavements with untreated foundations	S	Not recommended	Not recommended	May be used	May be used
	R	Not recommended	Not recommended	May be used in some cases	May be used
Pavements having foundations treated with hydraulic binders	S	May be used in some cases	May be used	May be used	May be used
	R	May be used in some cases	May be used in some cases	May be used in some cases	May be used
Cement concrete pavements	S	May be used in some cases	May be used in some cases	May be used	May be used
	P	Not recommended	Not recommended	May be used	May be used

Key: S= Subbase P= Plain concrete pavement R= Roadbase

Not recommended (dark shading) — May be used (blank)

May be used in some cases, depending on origin of concrete (pavements or buildings) (half-shaded triangle)

*Class T2: 150 to 300 5 ton axles per day } in most heavily
Class T3: 50 to 150 5 ton axles per day } travelled lane

However, "the possibility of using materials not in accordance with the Directives and Recommendations in force" is accepted for the design of pavements for low traffic levels (< T3), "because, in work of this type, and attempt is often made to use local materials."

Naturally, this exception has been used. A number of roads intended for low traffic levels have been built with foundations of untreated 0/30 to 0/40mm crushed-concrete aggregates, one of the oldest being the access roads to the Port de Créteil development zone in the Val-de-Marne department.

Since 1982, there have also been examples of foundations treated with hydraulic binders in pavements intended for higher traffic levels (T1 and T2) that have delivered altogether satisfactory mechanical performance, such as the F6 expressway, the RN 6 national highway in the Seine-et-Marne department (overlay), and the CD 60 departmental road in the Val-de-Marne department.

Other possible uses are being sought in mixtures with natural aggregates :

- the use of crushed-concrete aggregates in aggregate-slag mixtures to increase the crushing index
- the use of crushed-concrete sand in sand-slag mixtures to improve the stability of the formulations
- the construction of pedestrian ways, cycle paths and surfaces for sports.

6. Overall conclusions

The production of aggregates from demolition concrete is now operational. The quality of crushed-concrete aggregates has been substantially improved since 1982 thanks to technical advances in the plants and the general design principles of which now seem to have been well mastered. This should make it possible to simplify "approval" procedures for new installations.

The experience acquired since 1982 in the use of these crushed-concrete aggregates in road foundations would seem to be sufficient to confirm the possibility of wider use, in particular in pavements intended for low traffic levels.

References

Bauchard M., J-P. Joubert, Granulats élaborés par concassage de béton de démolition., Premier symposium international sur les granulats Nice, France, 21-23 mai 1984.

Buck A.D., Recycled concrete as a source of aggregate. Proc. of symposium, energy and resource conservation in the cement and concrete industry, Canmet, Ottawa 1976.

Darcel H., Verhee H., Chauchot J., Une opération innovante économe en énergie. La construction d'une voie express F6 en Essonne. Revue Générale des Routes et Aérodromes n°538, février 1982.

Frondistou-Yannas S. and Ng H.T.S., Use of concrete demolition waste aggregate in areas that have suffered destruction. A feasibility study., Report R 7737, Department of civil engineering, Massachusetts Institute of Technology, Cambridge MA nov 1977 (NTIS n° OS275888/AS).

Malhotra V.M., The use of recycled concrete as a new aggregate, Proc. of symposium energy and resource conservation in the cement and concrete industry, Canmet Ottawa, 1976.

Tavernier J., Riou R., Ray M., La réfection de l'autoroute du Nord entre Le Bourget et Roissy-en-France, chaussée province-Paris, Revue Générale des Routes et Aérodromes, n°531, mai 1977

Tomasawa F., Studies on the re-use of demolished concrete, Committee for research on the re-use on construction waste, Building contractors society, Tokyo, march 1975.

Wilson D.G., Foley P., Wiesman R., Frondistou-Yannas S., Demolition debris : quantities composition and possibilities for recycling, proceedings, 5th Mineral utilization symposium (Chicago, april 76) Aleshin, E. editor U.S. Bureau of Mines, Chicago IL, 1976, p.8-16.

Utilisation des déchets et sous-produits en technique routière, Rapport OCDE, Paris, 1977.

REUSE OF RECYCLED CONCRETE AGGREGATE FOR PAVEMENT

M. KAWAMURA Department of Civil Engineering
Kanazawa University
K. TORII Department of Civil Engineering
Kanazawa University

Abstract

From the viewpoint of energy and resources saving, it may be very advantageous to use the waste concrete as construction materials. Recycled aggregates obtained from old pavements can be effectively used as an aggregate for paving concrete or lean concrete base. This study mainly consists of two experiment. The experiment I aims at investigating the properties of recycled aggregates obtained from old pavements in relation to their application as an aggregate for paving concrete. The experiment II is concerned with the properties of high fly ash lean concrete made with recycled aggregates and its applicability as a material for base course. The experimental results show that recycled aggregates obtained from old pavements have more favorable physical properties as a concrete aggregate compared with recycled aggregates obtained from old buildings, and that recycled aggregates from pavement can be used as a concrete aggregate not only for lean concrete base but also for concrete paving.
Key words: Recycled aggregate, Concrete paving, Lean concrete base, Drying shrinkage, Freezing-thawing resistibility

1. Introduction

It may be significant to recycle the waste concrete produced from the construction of a new highway as a concrete aggregate for pavement from the viewpoint of solving such problems as solid waste disposal, reduction in construction cost and shortage of natural aggregates. In fact, the recycled aggregate from old concrete paving has been used as materials for pavement in the United States and European countries (Portland Cement Association 1980). However, recycled aggregates have not been actively used as a material for pavement in Japan (Kawamura and Torii 1984, 1986).

The objective of this study is to investigate the applicability of recycled aggregates as a concrete aggregate for paving concrete or lean concrete base course. The strengths, fatigue flexural strength, drying shrinkage and freezing-thawing resistibility of concrete made with recycled aggregates were investigated. Some properties of fresh and hardened high fly ash lean concrete made with recycled aggregates were also prepared in this paper.

2. Outlines of experiment

Three types of recycled aggregates are produced by crushing concrete rubbles obtained from old pavements and buildings by means of a jaw crusher ; Type A (strength of original concrete from pavement:370 kgf/cm^2, maximum size of coarse aggregate:40 mm), Type B (strength of original concrete from pavement:430 kgf/cm^2, maximum size of coarse aggregate:40 mm) and Type C (strength of original concrete from building:350 kgf/cm^2, maximum size of coarse aggregate:25 mm). Natural aggregates used are river sand (specific gravity:2.67, absorption capacity:1.3 %, finess modulus:2.73) and crushed stone (specific gravity:2.69, absorption capacity:0.8 %, finess modulus:7.13) from Hayatsuki river in Toyama Prefecture. The physical properties of recycled aggregates used are presented in Table 1. The combinations of fine and coarse aggregates in the preparation of specimens are river sand-crushed stone (Ns-Ng), river sand-recycled coarse aggregate (Ns-Cg) and recycled fine aggregate-recycled coarse aggregate (Cs-Cg). The ordinary Portland cement was used for making concretes and resin type Vinsol was used for entraining air in concrete. Two types of fly ashes produced in Takasago power station were also used for preparing high fly ash lean concrete specimens. The chemical and physical properties of fly ashes are presented in Table 2.

Table 1. Physical properties of recycled aggregates

	Specific Gravity	Absorption Capacity %	Abrasion Loss %	Crushing Value %	Finess Modulus
Recycled Fine Aggregate					
Type A	2.38	7.8	—	—	4.06
Type B	2.36	9.2	—	—	3.38
Type C	2.31	10.9	—	—	3.23
Recycled Coarse Aggregate					
Type A	2.45	5.4	24.5	24.4	6.92
Type B	2.48	4.9	34.8	23.6	6.96
Type C	2.42	5.9	29.5	29.2	6.94

Table 2. Chemical and physical properties of fly ashes

	Moisture Content %	Loss of Ignition %	SiO_2 Content %	Specific Gravity	Blain Specific Surface cm^2/g	Residual of 88μ Sieve %
Fly Ash						
Type A	0.2	1.9	54.7	2.27	3390	0.3
Type B	0.2	3.2	48.5	2.21	3140	11.4

The mix proportions of concretes for pavement in experiment I are as follows ; recycled aggregate used : Type A and C, cement content : 320-360 kg/m^3, water-cement ratio : 42 %, unit volume of coarse aggregate : 0.61-0.74. The unit water content and AE admixture dosage in concretes for pavement were determined so as to obtain the concretes with the slump of 2 cm and with the air content of 4 % according to the manual of cement concrete paving issued by Japan Road Association (Japan Road Association 1981). In experiment I , measurements were carried out on slump, V.B. value, compressive, tensile and flexural strength (JIS A 1108, 1113), flexural fatigue strength (minimum stress level : 10 % of static flexural strength, repeated load : 300 cycles/ minute), drying shrinkage (JIS A 1129) and freezing-thawing resistibility (ASTM C-666 A). All specimens were cured in water at 20 °C for a prescribed period.

The mix proportions of high fly ash lean concrete in experiment II are as follows ; recycled aggregate used : Type B, cement plus fly ash content : 200 kg/m^3, replacement by fly ash : 30, 50 and 70 %. The unit water content and AE admixture dosage in lean concretes were determined so as to obtain the concretes with the slump of 3 cm and with the air content of 7 %, which are requirements in the code of slipformed concrete (Portland Cement Association 1980). In experiment II , measurements were carried out on slump, bleeding capacity (JIS A 1123), compressive, tensile and flexural strengths, modulus of elasticity, pulse verocity, drying shrinkage and freezing-thawing resistibility.

3. Recycled aggregate concretes for paving concrete

3.1 Properties of the recycled aggregates and their influence on the properties of fresh concrete

As shown in Fig. 1, recycled aggregate grains are angular in shape and porous as a whole because of the mortars stuck on the surfaces of original aggregate grains. Recycled aggregates have low specific gravity and high absorption capacity compared with natural aggregates, as shown in Table 1. Judging from physical properties of the recycled aggregates, the recycled aggregate from an old pavement (Type A) is

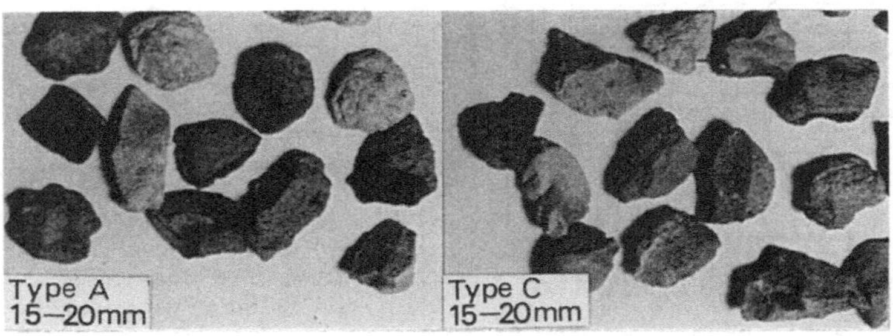

Fig. 1. Shape of recycled coarse aggregates

superior to that from a old building (Type C) as a concrete aggregate for pavement. The angularity of recycled aggregate grains influences consistency of concretes. For this reason, recycled aggregate concrete with Ns-Cg requires a greater unit water content by 10 to 15 % to obtain the same slump or V.B. value than that of natural aggregate concrete. Especially, the consistency of recycled aggregate concrete with Cs-Cg is considerably low and its ability of air entrainment is inferior.

3.2 Strengths of recycled aggregate concrete

Compressive, tensile and flexural strengths of recycled aggregate concrete are presented in Table 3. Compressive, tensile and flexural strengths of recycled aggregate concrete with Ns-Cg are somewhat lower than those of natural aggregate concrete. However, little difference in strength between the recycled aggregate concretes made with Type A and Type C was found. The influence of physical properties of recycled aggregate on the strength of concrete may not be so significant.

The flexural fatigue life of recycled aggregate concrete with Ns-Cg shows a great scatter compared with that of natural aggregate concrete.

Table 3 Compressive, tensile and flexural strengths of recycled aggregate concrete

	Strength kgf/cm^2				
	Compressive			Tensile	Flexural
	7 day	28day	90day	28day	28day
Ns—Ng	324	393	464	37	62
Ns—Cg Type A	305	378	443	35	60
Ns—Cg Type C	310	360	454	34	58

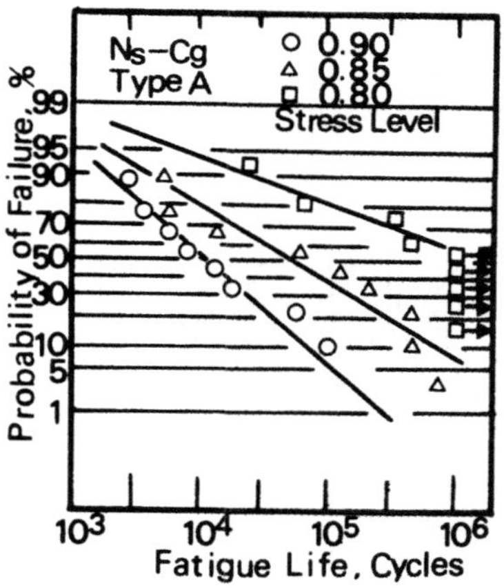

Fig. 2. Relationship between probability of failure and fatigue life

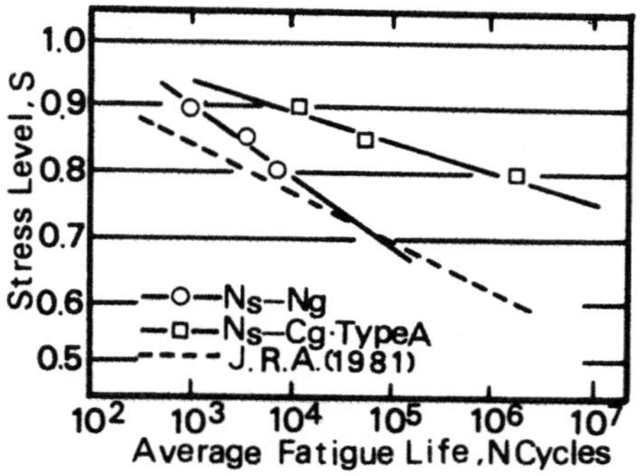

Fig. 3. S-N curves of recycled aggregate concrete and natural aggregate concrete

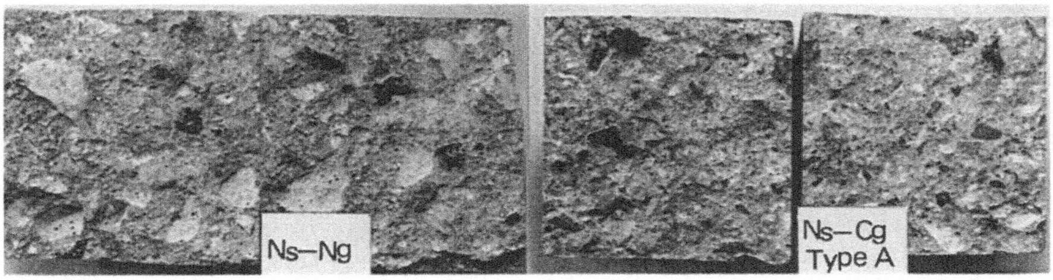

Fig. 4. Fracture surfaces of natural aggregate concrete and recycled aggregate concrete in flexural fatigue failure

Plotting the probabilities of failure against fatigue lives on logarithmic normal probability cordinates gives a linear relation between them, as shown in Fig. 2. The relationship between stress level (S) and average fatigue life (N) of recycled aggregate concrete with Ns-Cg (Type A) in flexural fatigue test is expressed as log N = 23.4-21.6 S. This equation gives the flexural fatigue strength of 78 % for 2 million cycles. The flexural fatigue strength of recycled aggregate concrete was found to be higher than that of natural aggregate concrete, as shown in Fig. 3. From a megascopic observation of fracture surfaces of specimens (Fig. 4), it was found that most of the failures in natural aggregate concrete occurred along the interface between cement mortars and aggregate grains, while in recycled aggregate concrete failures occurred within cement mortar portions of recycled coarse aggregate grains (Murdock 1957). Taking into considerations the difference in failure process between both, it can be concluded that a high flexural fatigue strength in the recycled aggregate concrete is derived from the strong bond between cement mortar matrix and recycled aggregate grains.

3.3 Drying shrinkage and freezing-thawing resistibility of recycled aggregate concrete

Drying shrinkage strains of recycled aggregate concretes are presented in Table 4. The recycled aggregate concretes show a greater drying shrinkage than the natural aggregate concrete. On the other hand, drying shrinkage of recycled aggregate concrete from pavement (Type A) is less than that of recycled aggregate concrete from building (Type C). The large drying shrinkage in recycled aggregate concrete may be resposible for the high water content of the concretes and the low

Table 4. Drying shrinkage of recycled aggregate concretes

	Drying Shrinkage		microstrain
	7 day	28day	90day
Ns–Ng	80	380	430
Ns–Cg TypeA	150	520	710
Ns–Cg TypeC	150	480	620
Cs–Cg TypeA	180	780	1050
Cs–Cg TypeC	160	630	940

modulus of elasticity of recycled aggregates. Furthermore, a relatively large drying shrinkage of recycled aggregate concrete with Cs-Cg appears to result from the high amount of mortar in the mixture.

The results of freezing-thawing repetition tests of recycled aggregate concretes are given in Table 5. Little deterioration was observed in both the natural aggregate concrete and recycled aggregate concrete with Ns-Cg (Type A), while the reduction in dynamic modulus of elasticity of the recycled aggregate concrete with Ns-Cg (Type C) and Cs-Cg (Type A and Type C) began to occur at early freezing-thawing cycles, having failed at under 300 cycles. Observations on the fracture surface of recycled aggregate concrete specimens with Ns-Cg (Type C) indicate that deterioration occurred along the interface between cement mortars and original aggregate grains or within cement mortars stuck on original aggregate grains. Therefore, in order to improve the resistance of recycled aggregate concrete against freezing-thawing repetitions, it is necessary to remove cement mortars stuck on original aggregate grains as far as possible.

Table 5. Freezing-thawing resistibility of recycled aggregate concretes

	Cycles to Failure	Durability Factor %
Ns–Ng	not failed	95
Ns–Cg·Type A	not failed	94
Ns–Cg·Type C	270	64
Cs–Cg·Type A	210	54
Cs–Cg·Type C	120	32

4. High fly ash lean concrete made with recycled aggregate (Type B)

4.1 Properties of fresh high fly ash lean concrete

The unit water content of high fly ash lean concrete decreases with the increasing percentage of replacement by fly ash, being 10 to 20 % less than the mixture without fly ash. The amounts of water required in recycled aggregate concretes are higher by 8 to 15 % than that in natural aggregate concretes. The airentaining ability of high fly ash lean concrete is mainly related to the characteristics of fly ash used and the percentage of replacement by fly ash, but the use of recycled aggregate does not influence the air-entraining ability.

The results of bleeding tests in high fly ash lean concrete are shown in Table 6. The bleeding of lean concrete is generally higher than that of normal concrete. The bleeding capacity of high fly ash lean concrete is affected by the type of fly ash used, the

Table 6. Bleeding capacity of high fly ash lean concretes

	Bleeding %	Final Time hours
Ns–Ng Plain	4.8	6:00
Ns–Ng FA·A 50	8.8	7:30
Ns–Ng FA·B 50	7.4	7:00
Ns–Cg Plain	7.6	7:30
Ns–Cg FA·A 50	9.9	11:00
Ns–Cg FA·B 50	9.6	10:30
Cs–Cg Plain	4.2	7:00
Cs–Cg FA·A 50	4.4	6:00
Cs–Cg FA·B 50	4.5	6:00

percentage of replacement by fly ash and the combinations of aggregate used. The addition of fly ash reduces the amounts of water required and delays the setting time, and thus decreases and increases the bleeding capacity of the mixtures, respectively (Ravina 1986).

4.2 Strengths of high fly ash lean concrete

Compressive strength, modulus of elasticity and pulse velocity of high fly ash lean concrete made with recycled aggregate are presented in Table 7. Although the compressive strength of high fly ash lean concrete is smaller than that of lean concrete without fly ash at early ages, high fly ash lean concrete shows continuous strength development after 28 days (Harque 1984). The degree of strength development in high fly ash lean concrete is remarkable as the percentage of replacement by fly ash increases. On the other hand, the influence of the use of recycled aggregate on the compressive strength of lean concrete seems to be not so significant as in normal concrete

Table 7. Compressive strength, modulus of elasticity and pulse velocity of high fly ash lean concretes

	Compressive Strength kgf/cm2			Estatic 10⁴·kgf/cm2	Edynamic 10⁴·kgf/cm2	Vpulse m/sec
	7 day	28day	90day	28day	28day	28day
Ns—Ng Plain	119	184	219	21	29	4260
Ns—Ng FA·A50	60	92	167	15	26	4030
Ns—Ng FA·B50	49	79	147	10	23	4000
Ns—Cg Plain	96	134	185	21	24	4060
Ns—Cg FA·A50	55	109	169	19	22	3950
Ns—Cg FA·B50	42	91	127	18	24	3740
Cs—Cg Plain	109	157	176	19	24	3270
Cs—Cg FA·A50	29	72	106	15	19	3740
Cs—Cg FA·B50	43	87	138	17	21	3590

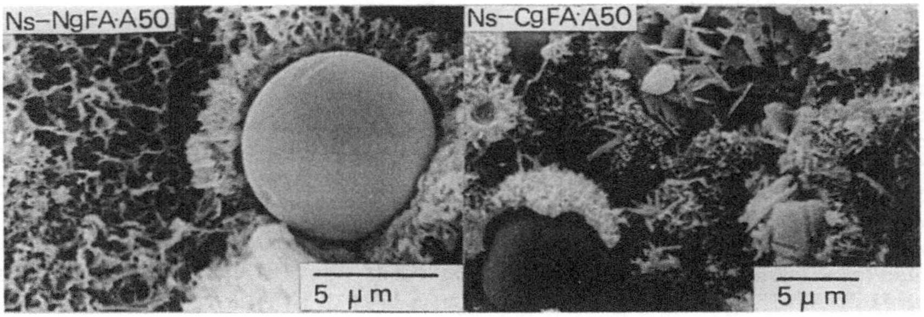

Fig. 5. Scanning electron micrographs of fracture surfaces of high fly ash lean concretes (F/C+F=50 %, 90 days)

(Kawamura 1983). As shown in Fig. 5, the pozzolanic reaction products such as reticulated network of C-S-H gel, platy C-A-H and needle-like ettringite were formed in the vicinity of fly ash particles during the long-term curing in water.

The relationship between compressive strength and dynamic modulus of elasticity in high fly ash lean concrete is shown in Fig. 6. The dynamic modulus of elasticity in high fly ash lean concrete proportionaly increases with an increase of compressive strength in all combinations of aggregates used. However, high fly ash lean concrete made with the recycled aggregate shows a little smaller modulus of elasticity than natural aggregate concrete. These results indicate that high fly ash lean concretes with the replacement up to 50 % can be used enough as a base course material in all combinations of the recycled aggregates.

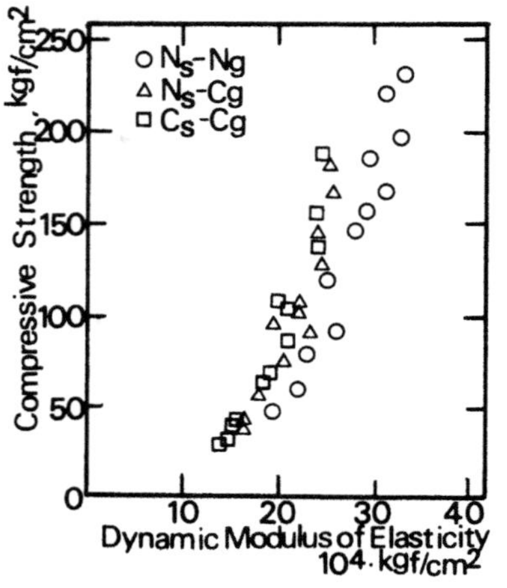

Fig. 6. Relationship between compressive strength and modulus of elasticity in high fly ash lean concretes

4.3 Drying shrinkage and freezing-thawing resistibility of high fly ash lean concrete

Drying shrinkage strain of high fly ash lean concrete made with recycled aggregate is presented in Table 8. Drying shrinkage in high fly ash lean concrete at early ages is a little greater than that in lean concrete without fly ash. However, the rate of drying shrinkage in high fly ash lean concrete after the age of 28 days significantly decreases as the percentage of replacement by fly ash increases in all combinations of aggregate used. The average value of drying shrinkage in high fly ash lean concrete made with natural aggregates is approximately 500 microstrains at the age of 90 days. Drying shrinkage in high fly ash lean concrete made with recycled aggregates is 1.2 to 1.5 times larger than that of high fly ash lean concrete made with natural aggregates. The low drying shrinkage of high fly ash lean concrete is beneficial for preventing reflection cracks in pavement.

Table 8. Drying shrinkage of high fly ash lean concretes

	Drying Shrinkage microstrain		
	7 day	28day	90day
Ns —Ng Plain	90	460	710
Ns —Ng FA·A50	120	380	450
Ns — Cg Plain	110	540	690
Ns —Cg FA·A50	110	510	620
Cs —Cg Plain	160	680	990
Cs —Cg FA·A50	140	650	880

The results of freezing-thawing tests of high fly

ash lean concrete made with recycled aggregates are shown in Table 9. Freezing-thawing resistibility of high fly ash lean concrete with natural aggregates is not so greatly different from that of lean concrete without fly ash as far as the percentage of replacement by fly ash is within 50 %. On the other hand, the influence of recycled aggregates on the resistibility against freezing-thawing repetitions is significant. The resistibility of high fly ash lean concrete against freezing-thawing repetitions is smaller in the following order of combinations of aggregate ; Ns-Ng, Ns-Cg, Cs-Cg. However, usually the lean concrete used as a base course material will not be exposed to the severe freezing-thawing repetitions except for the portion of pavement shoulder.

Table 9. Freezing-thawing resistibility of high fly ash lean concretes

	Weight Loss %	Durability Factor %
N_s-Ng Plain	5.4	87
N_s-Ng FA·A 50	6.5	84
N_s-Cg Plain	4.8	48
N_s-Cg FA·A 50	9.2	40
Cs-Cg Plain	4.7	52
Cs-Cg FA·A 50	8.5	40

5. Conclusions

The purpose of this study is to clarify the applicability of recycled aggregate to paving concrete and lean concrete base course. It was found that the recycled aggregate from an old pavement had more favorable physical properties as a concrete aggregate for concrete paving than that from an old building. Judging from the mechanical properties of the concretes made with the recycled aggregate, the recycled aggregate from the old pavement can be used as a concrete aggregate for the light-traffic pavement or car parking. However, it should be noted that the recycled aggregate concretes show a high drying shrinkage and considerably low freezing-thawing resistibility. It was also found that the high fly ash lean concrete made with recycled aggregates was prospective as a material for the base course because of its good workability, continuous strength development, high rigidity and low drying shrinkage.

References

Portland Cement Association (1980) Lean concrete (econocrete) base for pavements : Current practices, Concrete Information, U.S.A..

Torii, K. and Kawamura, M. (1984) Applicability of recycled concrete aggregate as an aggregate for concrete pavement, Trans of the Japan Concrete Institute, 6, 133-140.

Torii, K. and Kawamura, M. (1986) Application of high fly ash lean concretes to the subbase for pavement, Trans. of the Japan Concrete Institute, 8, 37-44.

Japan Road Association (1981), Manual of Cement Concrete Pavement (in Japanese).

Murdock, J.W. and Kesler, C.E. (1957) Effect of range stress on fatigue strength of plain concrete beams, J. Am. Concrete Inst., 30, 2, 222-231.

Ravina, D. and Mehta, P.K. (1986) Properties of fresh concrete containing large amounts of fly ash, Cement and Concrete Research, 16, 227-238.

Harque, M.N., Langan, B.W. and Ward, M.A. (1984) High fly ash concretes , J. Am. Concrete Inst., 81-8, 54-60.

Kawamura, M. and Torii, K. (1983) Properties of recycling concrete made with aggregates obtained from demolished concrete pavement, J. of Materials Science, 32, 353, 208-214 (in Japanese).

REUSE OF CRUSHED CONCRETE AS A ROAD BASE MATERIAL

D. GORLE, L. SAEYS
Belgian Road Research Center, Brussels

Abstract
An extensive study has been undertaken on 7 test roads and, in the laboratory, on 84 samples coming from 10 different Belgian crushing installations, to evaluate crushed concrete, in comparison with classical stone, as a road base material.
To produce a good crushed concrete aggregate, measures are to be taken already during the demolition of the concrete pavement : sufficient reduction of the concrete slabs, selective excavation, separation of the fine fraction before the crusher, and continuous maintenance of the crusher installation to obtain a grain size distribution as for classical base materials.
The density of the compacted material will be lower than for classical crushed stone, due to the internal voids of the concrete. The bearing capacity is often lower immediately after compaction, but may improve and even become superior because of the residual cement liberated during crushing.
The recycling of crushed concrete can be economically very advantageous on account of the reduction in transport costs and the avoidance of dumping costs. It has been shown that a reduction of the maximum size of the crushed concrete to the classical limits will increase only slightly the total cost and has important technical advantages.

Key words : concrete, crushing, recycling, road base.

1. Introduction

The amounts of demolition concrete are increasing every year and so is the necessity to reuse these materials for economical and ecological reasons, especially in densely populated areas. Properties of recycled aggregates and methods to reuse in concrete or in base and subbase layers of roads have been described (see References). In this paper the reuse of aggregates from crushed concrete is examined. The production of the aggregates, the laboratory study, the test roads and the economical aspects are described successively.

2. Production of crushed concrete aggregates

In order to produce, with a sufficiently high output, a good aggregate from demolition road concrete, special measures are to be taken already during the demolition of the pavement. On the road site the concrete slabs have to be reduced, most often by a hydraulic hammer, to dimensions adapted to the selected crusher. This implies in many cases a higher degree of reduction than what would be required if the concrete is not recycled. The excavation of the demolished concrete has to be selective in order to avoid pollution by e.g. fine soils, which is important for the frost susceptibility and sensitivity to water. For the same reason removal of the fines ahead of the crusher is necessary. This is normally achieved by the grizzly feeder.

If reuse of the crushed concrete as a base material on the same site is possible, then the most economical solution (see § 5) is to install a small mobile crusher (Fig. 1) on the construction site and to move this installation along as the road works progress, to limit the transportation costs. The use of a conveyor to feed the crusher allows to improve the mobility of the installation by avoiding the construction of an embankment ahead of the crusher. This type of installation can produce in one operation, without screening or recycling, a continuous grain size distribution suitable for unbound road bases as shown in Fig. 2. This implies a continuous maintenance and adjustment of the crusher since wear of the jaws has an important influence on the grain size.

3. Laboratory study of crushed concrete

Classical laboratory tests, on 84 samples of crushed concrete aggregates from about 10 different Belgian crusher installations, have shown that most of these materials are suitable for road bases or subbases according to the Belgian specifications. Some of the materials however are too coarse which causes problems during construction (see § 4).

The specific density of crushed concrete aggregates is lower than for classical materials which has to be taken into account in the specifications for the density of road bases or subbases using these materials.

The crushed concrete aggregates have been compared with the classical materials for bases and subbases, from the point of view of deformation under repeated loading. The triaxial test set up is shown in Fig. 3, with a typical result for the total vertical relative deformations, the permanent deformations and the reversible deformations (Fig. 4). These deformations are measured on the central part (2/3) of the sample in order to avoid the disturbing effects of the bottom

Photo CRR S/2695.

Fig. 1 Mobile jaw crusher with conveyor for feeding.

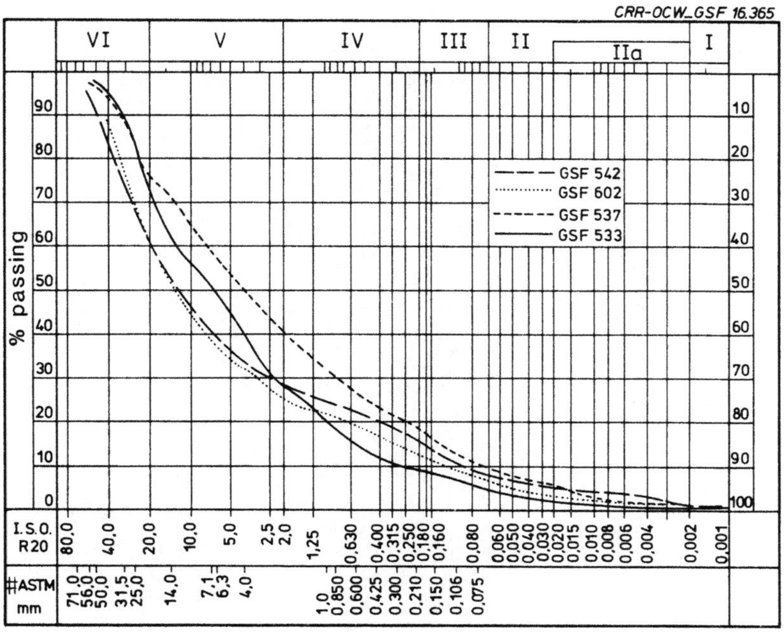

Fig. 2 Grain size distribution of crushed concrete (jaw crusher without recycling)

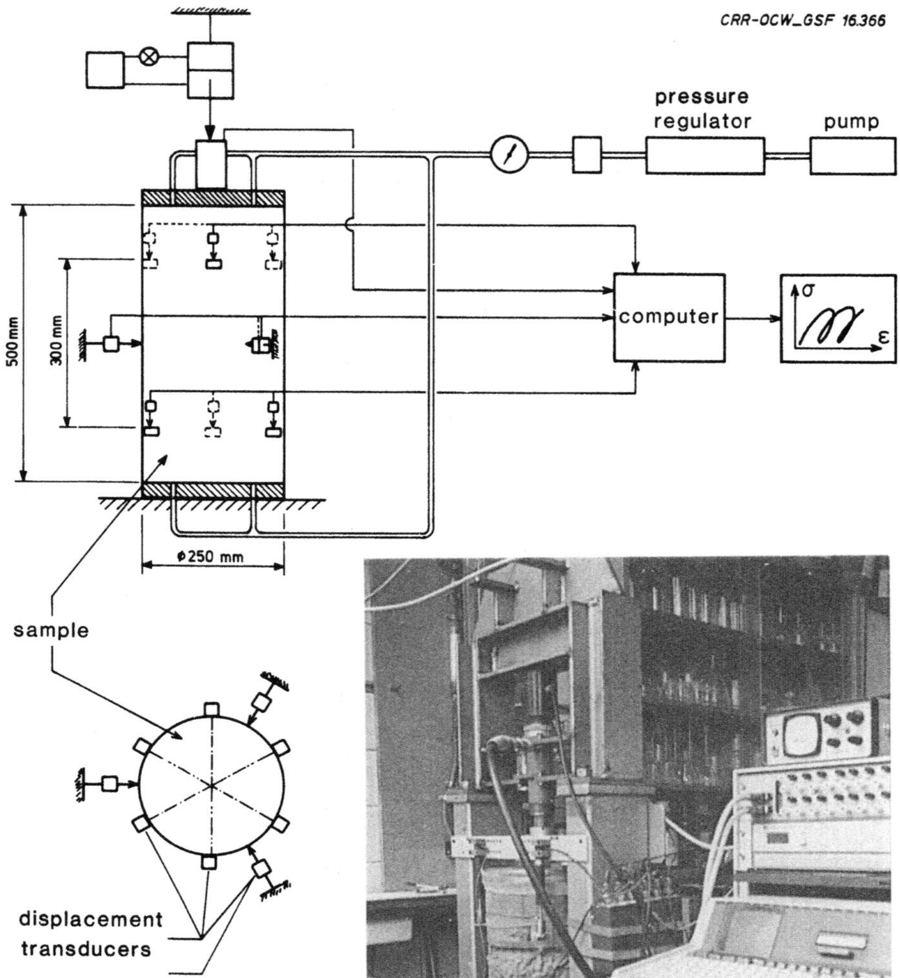

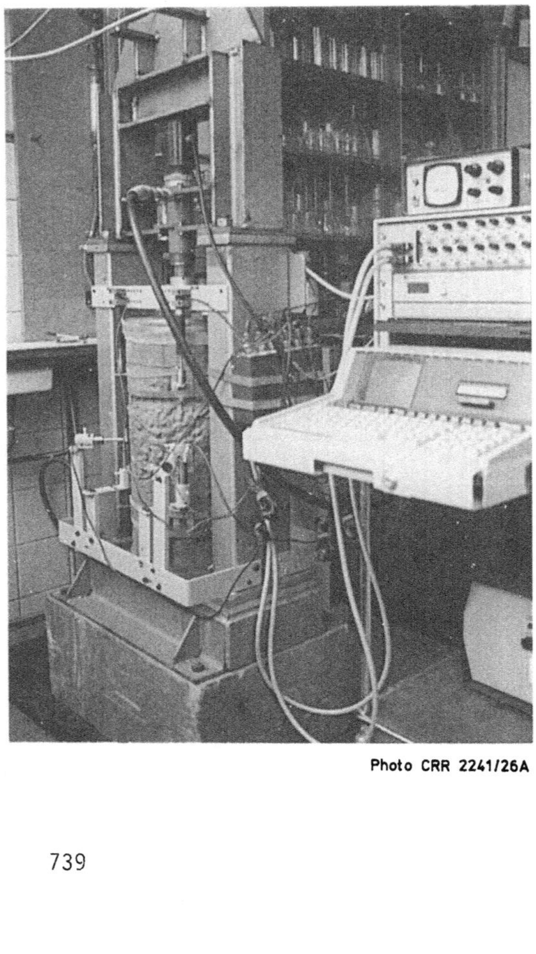

Fig. 3 Triaxial test setup

Photo CRR 2241/26A

and top plate. The total relative vertical deformations can be written as (Descornet 1978) :

$$\epsilon_{zt}(N) = \epsilon_{zt}(1) \cdot N^{\alpha}$$

$$\alpha = S \cdot \frac{\epsilon_r}{\sigma_o}$$

where N is the number of loads
α an exponent
S the susceptibility to deformation
ϵ reversible deformation
σ_o confining stress.

The elasticity modulus, calculated after a sufficiently high number of cycles to obtain a constant reversible deformation and thus a constant value for the modulus, is given in Fig. 5 as a function of the confining stress and compared with the classical materials. It appears that the modulus of the crushed concrete is lower than those for the others materials but higher than the modulus for sand. Fig. 6 shows that the crushed concrete behaves better than the other materials from the point of view of permanent deformations.

4. Test roads

Crushed concrete aggregates have been used in the base or subbase of 7 test roads. The observation and experiments on these roads lead to the following conclusions :

- some of the materials have a too large percentage of grains larger than the specified maximum size for classical materials. This gives several problems during construction : segregation, too large dispersion of the density and uneven surface etc. For these reasons, it is very important to respect the maximum grain size, and this can be achieved with only a slight increase of the total costs.

- the average density is much lower than what is normally required for classical materials (Fig. 7), this is due to the lower specific density of crushed concrete aggregates (see § 3),

- repeated plate bearing tests, immediately after construction, show that the moduli of the layers with crushed concrete are on the average lower than those of the classical materials : approximately 410 MPa for crushed concrete compared to 680 MPa for the usual crushed stone. This confirms the laboratory measurements. Some of the crushed concrete layers show however a very important increase of the modulus in the period of several weeks after construction (Fig. 8). This can be explained by the residual cement present in crushed concrete.

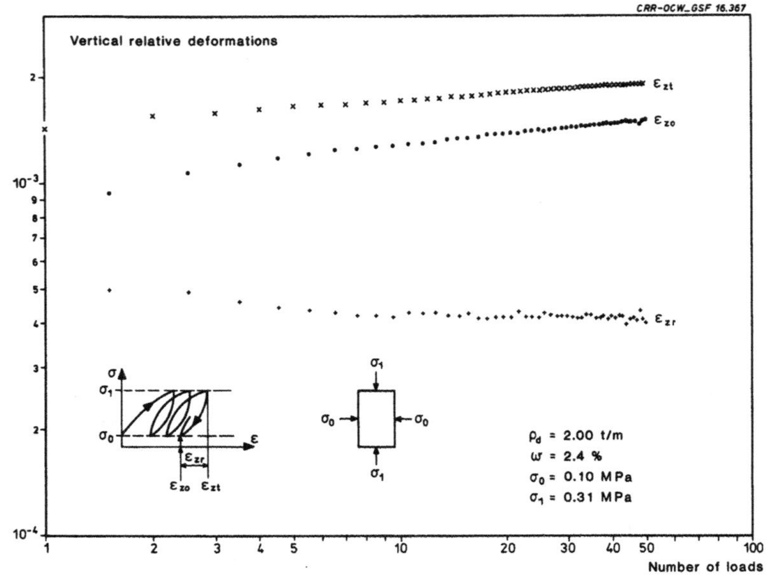

Fig. 4 Total vertical relative deformations (ε_{zt}) after unloading (ε_{zo}) and reversible (ε_{zr})

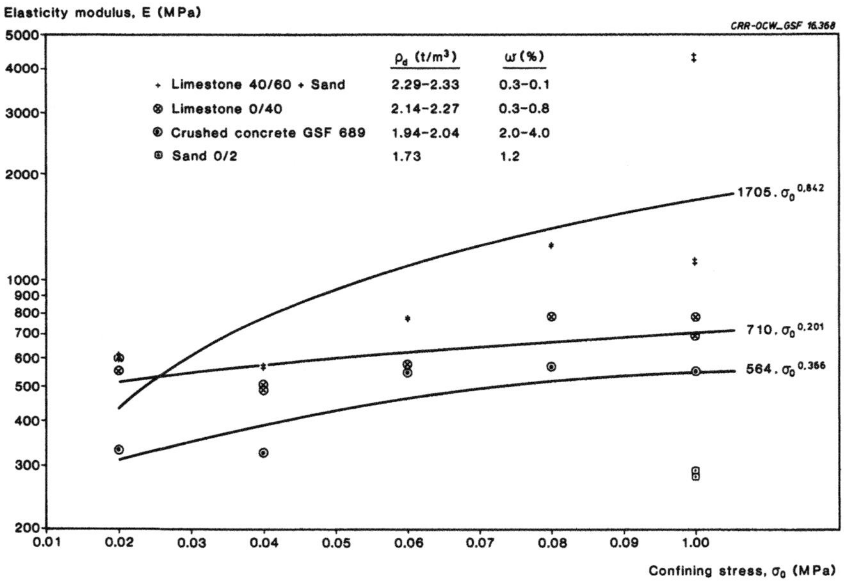

Fig. 5 Elasticity modulus versus confining stress

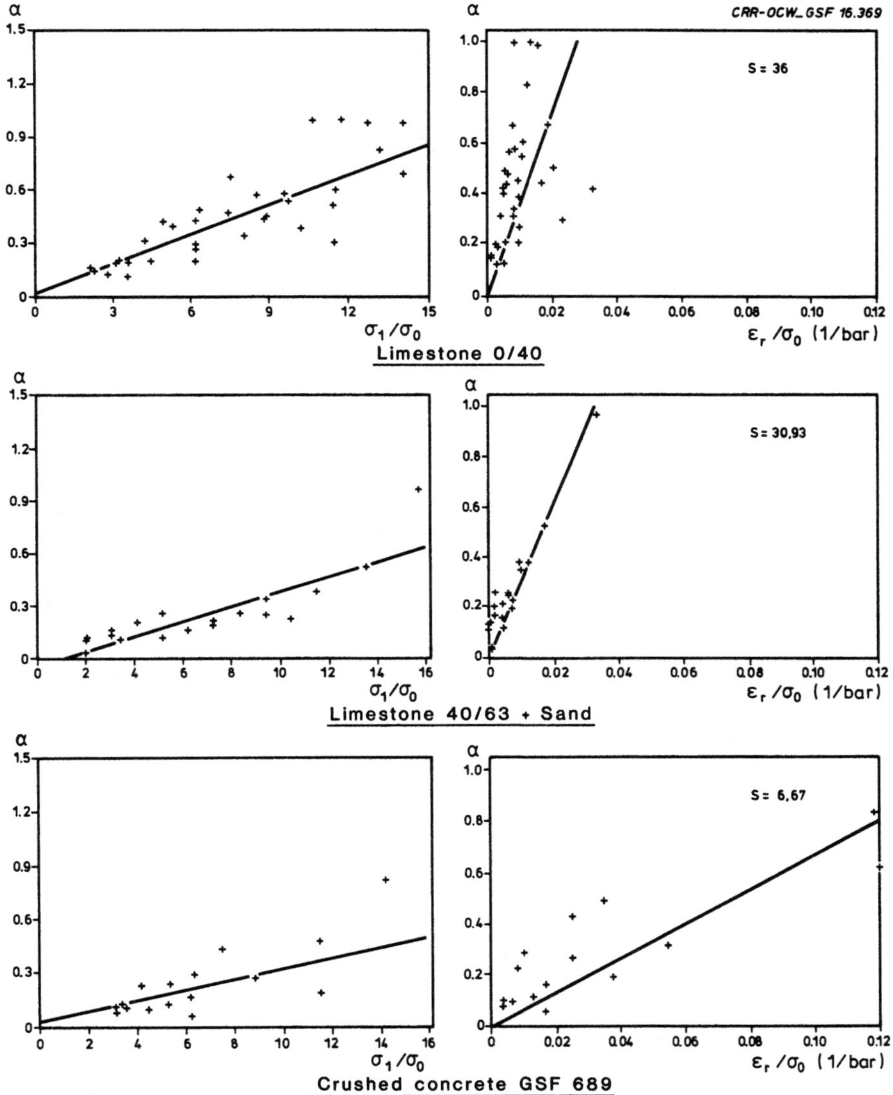

Fig. 6 Exponent α versus σ_1/σ_0 and versus the ratio of the reversible deformation (ε_r) to the confining stress (σ_0) for crushed concrete in comparison with classical materials

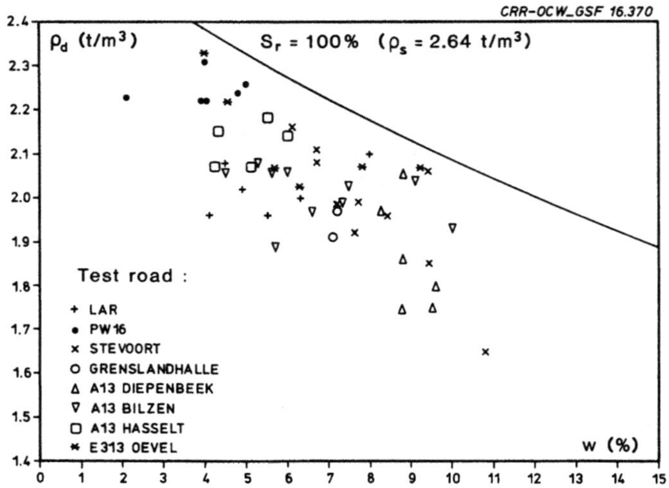

Fig. 7 Dry density versus water content of base layers of crushed concrete

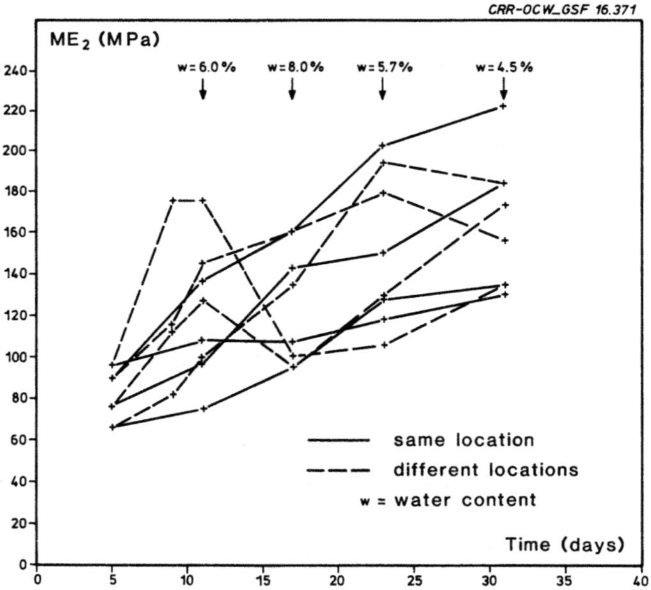

Fig. 8 Evolution of the compressibility modulus of the 2nd cycle of crushed concrete base layers

5. Economic aspects

Economically speaking, the recycling of crushed concrete from demolished plain concrete pavements is a very advantageous procedure, especially if crushing can be carried out on the site and if the aggregate obtained can be used in situ as an unbound road base or sub-base material.

Comparative costing of the conventional method of execution, which consists of transporting and discharging the demolition waste into the nearest dumping site and subsequently delivering new road base or sub-base materials along the road, and the above-described procedure for the in situ recycling of the concrete after crushing has shown that, for the total of operations including the complete demolition of the concrete pavement and the road base and the subsequent laying of the new road base, the cost of the latter is only 55 % of that of the former.

70 % of the total savings comes from reduced transport costs, 20 % from the smaller cost price of the material and 10 % from avoided dumping costs. In this comparison, the average distance from the job to the waste dump and the extraction site for the new materials is 15 and 35 km, respectively.

Sharing out the total savings per job item results in 70 % being assigned to the laying of the road and 30 % to the demolition of the pavement.

The comparative cost analysis has also revealed that breaking down the concrete to blocks of sufficiently small a size not only hardly changes the cost of the demolition operation, but also encourages the use of a mobile crusher while respecting the optimum ratio between the size of the waste concrete to be processed and that of the aggregate to be obtained.

Special attention must be paid to the feeding of the crusher, which accounts for 70 % of the total cost of the crushing operation ; the cost of sieving the crushed material and carrying it off by conveyor belts is relatively less important.

Acknowledgements

This research was carried out under the direction of M. J. Reichert, Director of the Belgian Road Research Center and M. J. Verstraeten, Chief of the Department of Research and Development, and with the financial help of the IRSIA (Institut pour l'Encouragement de la Recherche Scientifique dans l'Industrie et l'Agriculture) of Belgium.

References

Bauchard, M., Joubert, J.P., Granulats élaborés par concassage de bétons de démolition. Bulletin de liaison des laboratoires des Ponts et Chaussées, n° Spécial XIV, 150-153, Paris (1984).

Bernier, G., Marlier, Y., Mazars, J., "Un matériau nouveau provenant de la démolition du béton : le "bibéton" C.R. Conférence internationale". Sous-produits et déchets dans le Génie Civil. Vol. I, blz. 157-162, Association Amicale des Ingénieurs Anciens Elèves de l'Ecole Nationale des Ponts et Chaussées, Paris, 1978.

Buck, A.D., Recycled concrete, Highway Research Record No. 430, 1-8, Washington (1973).

Descornet, G., Personal information, Centre de Recherches Routières, Bruxelles, 1978.

Frondistou-Yannas, S., Taichi Itoh, Economic feasability of concrete recycling. Proc. American Society Civil Engineers, Journal of the Structural Division, Vol. 103, ST 4, blz. 885-899 (1977).

Maagdenberg, A.C., Hendriks, Ch. F., van Somselaar, M.J., Sweere, G.T.H., Granulaat van beton- en metselwerkpuin als ongebonden steenfunderingsmateriaal. Wegen, 57 nr. 12, blz. 392-397, Delft, 1983.

Merrien, P., Nissoux, J.L., Les enseignements de deux chantiers de reconstruction de l'autoroute du Nord. Bulletin de liaison des laboratoires des Ponts et Chaussées, n° 96, 95-109, Paris (1978).

Nat. Coop. Highway Research Program, Guidelines for recycling pavement materials, NCHRP 224, Washington, 1980.

Petrarca, R.W., Galdiero, V.A., Summary of Testing of Recycled Crushed Concrete, Transportation Research Record 989, 19-26, Washington, 1984.

Ray, G.K., Recycling Portland Cement Concrete. Portland Cement Association, Skokie, Illinois, 1979.

Service d'Etudes Techniques des Routes et Autoroutes, Réemploi de béton de démolition dans le domaine routier. SETRA, Note d'information, Paris, 1984.

Sommer, H., Recycling von Beton- Wiederverwendung im Deckenbau, Strasse und Autobahn 35, Bonn, 1984.

Van Heystraeten, G., Recyclage de débris de béton de construction routière, La Technique Routière XXIX/4, 8-31, Bruxelles, 1984.

REUSE OF CONCRETE PAVEMENTS

KNUD PUCKMAN National Road Laboratory, Denmark
ANDERS HENRICHSEN Dansk Vejbeton A/S, Denmark

Abstract:
Concrete road pavements, parking areas, etc at some stage deteriorate to such a degree that they have to be removed or replaced by new materials. When the time of replacement has been reached, the question arises of what to do with the old concrete which consists of some 80% sand and gravel and some 20% hardened cement. Can these materials be recycled as aggregate in the production of high quality concrete for new concrete pavements.

The study has included 4 different concrete pavement materials: a pavement at an airport from 1959, a road pavement from 1929 and two motorway pavements from 1968, of which one of the latter had deteriorated due to alkali silica reactions.

Where existing pavement has shown rapid deterioration due to alkali silica reaction and/or bad concrete quality, it is not advisable to recycle the materials in a new concrete pavement. Existing pavement deteriorated due to fatigue and excessive loading after many years of use should be eligible for recycling.
Key words: Concrete pavements, Recycling, Water absorption, Alkali-silica reactivity, Quality, Shrinkage.

1. Introduction

It was examined, whether existing concrete pavements which are worn down/deteriorated, can be recycled as aggregate in new, durable concrete pavements. It is expected that direct savings can be obtained, transportation requirements can be reduced, existing gravel resources can be saved and finally the problem of removal and storage of the large concrete slabs weighing some 10 tons is solved.

The tests have included:
- Description and analysis of the pavements before removing.
- Determination of the properties of the crushed concrete aggregates.
- Preparation of laboratory mixes in which crushed concrete has been used entirely or partially as aggregate as well as preparation of a reference concrete with natural aggregates. All mixes have an identical content of cement, microsilica, water and additives.
- Test of the properties of the fresh and hardened recycled concrete.
- 70 m long full-scale test stretch with recycled concrete at Copenhagen Airport, Kastrup.

2. Test objects

Special importance was attached to choosing concretes from test objects which differed from each other in composition, method of work and deterioration mechanisms.

KAS - Apron, Copenhagen Airport
The pavement was constructed in 1959 with traditional equipment. Thickness 0.30 m (2-layer concrete). Reinforcement 2.2 kg/m^2. Slab width 4.60 m. Transverse joints every 5 m. Cement content 300 kg/m^3. W/C ratio 0.50. Not air-entrained.

The pavement was removed in 1985 in connection with the rebuilding and reinforcement of the apron. The characteristic defects of the concrete pavement before breaking up were cracked and patched slabs, displaced slab joints, scaling and damage along the edges. Actual chemical deterioration (alkali silica reactions) or frost damage was not observed to any great extent.

CXA - Road pavement, Chr. X. Allé, Lyngby
The pavement was constructed in 1929 with a tamping machine. Thickness 0.12 m in the middle and 0.15 m at the edge. Reinforcement 5 kg/m^2. Slab width 6 m (2x3 m). Transverse joints every 17-25 m. Cement content 350 kg/m^3. W/c ratio 0.44. Not air entrained.

The pavement was covered with an asphaltic layer in 1953. In 1985/86 the pavement was removed due to unevenness, deterioration around the joints and cracks as well as partial rebuilding.

SMV - Southern Motorway Cordoza-Dyrehavehus
The pavement was constructed in 1968/69 with a slipform paver. Thickness of 0.20 m. Slab width 7.5 m (2 x 3.75 m). Transverse joints every 5 m. Cement content 320 kg/m^3. W/c-ratio 0.52. Air entrained. Spacing factor 0.25 mm.

The pavement was covered with an asphaltic layer in 1975 due to reduced friction and signs of deterioration of the concrete surface in the form of scaling.

MRV - Motor Ringroad, Buddinge-Ring III
The pavement was constructed in 1968 with a slipform paver. Thickness 0.20 m. Slab width 7.5 m (2 x 3.75 m). Transverse joints every 5 m. Cement content 320 kg/m^3. W/c-ratio 0.55. Air entrained. Spacing factor 0.12 mm.

The pavement was covered with an asphaltic layer in 1972 due to scaling and reduced friction. The entire pavement was removed in 1983 due to deterioration primarily from alkali silica reactions.

3. Examination of crushed materials

The materials were taken up by means of a hydraulic hammer and crushed with a cone/jaw crusher and then sieved to obtain a product with a maximum grain size of 32 mm. Representative samples were then taken to determine the initial properties of the recycled materials. Apart from sieve analysis, particle shape, mortar content in the crushed concrete fractions, content of flint/porous flint and chloride content - all the results can be found in [1] - water absorption and alkali silica reactions for the sand was examined.

Water absorption

The water absorption of the crushed concrete is markedly higher than for naturally occurring aggregate which is due to the fact that the mortar content of the crushed particles has a great ability for water absorption.

Absorption values for crushed concrete stated in literature are approx. 10 times greater than those for naturally occurring aggregates [2].

In view of the fact that water is an important factor in the production of any concrete, special attention has been paid to the determination of water absorption. It has been proved that the method used until now - International Standard Organization 6783 is not suitable for crushed concrete. It is almost impossible to determine whether the material has reached a condition which can be defined as water saturated, surface dry due to the shape and the rough/porous mortar surface of the crushed particles. It is due to this special condition that too high water absorption is measured when using this standard method.

An examination has therefore been made whether the water saturated, surface dry condition can be determined graphically by means of a desorption isotherm of a sample - a graph of drying at a chosen steady temperature. This drying curve can be divided into 4 characteristic parts as shown in fig. 1.

The 4 parts of the curvature consist of:

- A-B A steep part of the curve showing the evaporation of the free water on the particle surface
- B-C A curve where the water in the large voids evaporates
- C-D An almost horizontal part of the curve when the water in the small voids evaporates
- D-E A horizontal part of the curve when all the water that can evaporate has been removed.

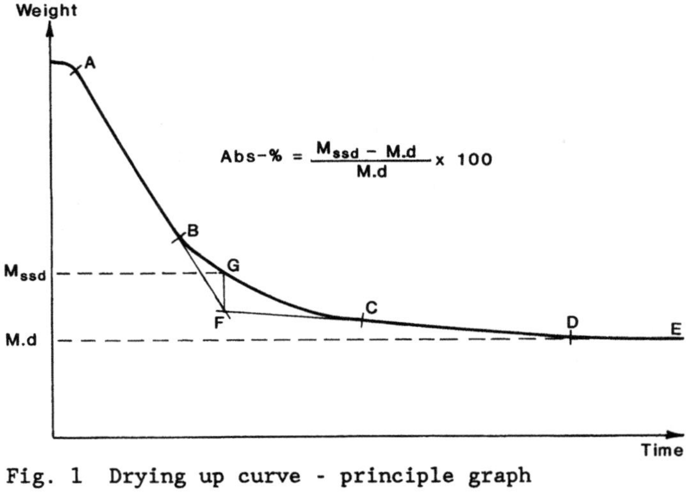

Fig. 1 Drying up curve - principle graph

It is assumed that the water saturated, surface dry condition will correspond to the point (G) on the curvature BC and that this point can only be established graphically by plotting the vertical line from the point (F) which is the cross point for the extension of the two linear parts of the curvature AB and CD.

The absorption of the sample can then be determined in the normal way by substituting the values for the points G and D.

Natural sand as well as crushed concrete aggregate has been tested and the results have been compared to measurements obtained on the basis of the ISO-method.

In order to support the graphic method, a determination of the water absorption of material samples by storage in a climate chamber was undertaken until equilibrium was obtained at approx. 100% relative humidity; a calculated evaluation of the materials water absorption potential was also carried out.

The results of the measurements using the 3 methods can be seen in table 1. A further description of the background and method for determining the graphic water absorption is given in [1].

The measurements show that water absorption of crushed concrete determined by the graphic method as well as by storing at 100% relative humidity is less than that obtained by means of the ISO-method.

A need has therefore arisen to develop a new standard test method for determination of water absorption of crushed concrete. Such a method could be based on desorption curves, provided it can be developed also to include particles greater than 4 mm when using larger test samples. This also requires that the method is verified with a round Robin test.

Material	Fraction mm	Abs-% by means of ISO	IR	100 RF
Sea sand		0.9	1.4	
Pit sand		1.9	1.5	
Crushed concrete	0-4			
KAS		7.8	3.6	2.9
CXA		7.4	3.6	2.7
SMV		7.9	3.8	3.9
MRV		11.4	4.5	3.6
KAS	4-8	7.1	2.7*	2.8
CXA		4.2	1.8*	1.9
SMV		6.2	2.7*	3.0
MRV		11.2	3.1*	3.1
KAS	8-25	4.7	0.6*	1.7
CXA		3.2	1.0*	1.6
SMV		4.0	1.5*	1.9
MRV		5.1	1.4*	1.8

Comments:
Average of 3 samples
* Absorption values could be incorrect due to small sample sizes.

ISO: International Standard Organization

IR: New graphic method, infrared drying

100RF: Storing to constant weight in 100% relative humidity

Table 1: Water absorption determined by using 3 different methods.

Testing of the alkali silica reactivity of crushed concrete sand

The sand fractions of crushed concrete (materials less than 4 mm) have been examined for any damaging alkali silica reactions by measuring the expansion of cast mortar prisms (40 x 40 x 160 mm) by storing them in saturated salt water (NaCl) at 50°C. Measurements are made after 8 and 20 weeks' storage.

In accordance with existing practice for the classification of sand for quality concrete, sand is considered alkali silica reactive if the expansion after 8 weeks is greater than 1 ‰. For road concrete this demand will normally be extended to apply also to storage after 20 weeks.

The result of the examination was - stated as an average of 3 measurements:

Crushed concrete sand (0-4 mm) from:	Expansion in ‰ 8 weeks	20 weeks
KAS	0.09	0.33
CXA	0.15	0.19
SMV	0.29	0.49
MRV	0.69	1.66

Table 2. Expansion of mortar bars.

Evaluation:
Crushed concrete sand from the Motor Ringroad is not suitable for pavement concrete. The other tested materials can be considered as suitable.

4. Determination of the properties of recycled concrete

A series of concretes mixes using the crushed materials with an identical content of cement (approx. 310 kg/m^3), microsilica (14-15 kg/m^3), water (w/c-ratio approx. 0.38) and additives (superplasticizer + air entraining 5-6%) was prepared. Mixes were prepared as well with the entire, crushed material (total) as with the 0-4 mm fraction removed (sieved), which then was replaced by natural sand. For the sake of comparison a reference concrete was mixed (named REF in the following) with the same mix design but with natural sand and crushed granite as aggregate.

Samples were taken/prepared from the mixes in order to determine the properties of both the fresh and the hardened concrete. 35 concrete cylinders 100/200 mm were cast from each type of concrete and 5 concrete bars 100x100x600 mm, corresponding to a total of 315 cylinders and 45 bars. The cast samples were stored under standard conditions and maturity exceeded 28 days before testing.

The following material combinations were used:

Test object	Mix no.	Crushed concrete		Natural materials	
		Sand 0-4 mm	Stone 4-32 mm	Sea sand 0-4 mm	Granite 4-32 mm
REF	0			x	x
KAS	3 5	x	x x	(x) x	
CXA	4 7	x	x x	(x) x	
SMV	1 6	x	x x	(x) x	
MRV	2 8	x	x x	(x) x	

Table 3: Combination of the material.

Comments:
(x) Since the crushed material only had a content of 24-30% of the 0-4 mm fraction, it was corrected by adding natural sand in order to obtain a concrete mix of the same consistency (slump 25-30 mm) as the other mixes.

5. Results

Compressive strength, tensile strength and the dynamic E-modulus (all average of 3 specimens - 100/200 mm cylinders) are shown in fig. 1.

The drying shrinkage was measured continuously on concrete bars 100 x 100 x 600 mm, stored at 20°C at 50% relative humidity.

The results are an average of 5 tests and are presented graphically in fig. 2.

The recycled concretes have a drying shrinkage which is from 39% (mix 5) to 114% (mix 8) greater than that of the reference concrete.

The frost resistance test was carried out in accordance with "Richtlinien zur Bestimmung und Prüfung der Frost-Tausalzbestandigheit von Zementbeton", worked out by Dubrolubov and Romer in 1977.

3 prisms 30 x 30 x 70 mm were sawed from the concrete cylinders prepared in the laboratory for each concrete mix; after 3 days' storage, they were exposed to 200 freezing and thawing cycles consisting of 20 mins in a freezing bath and thereafter 10 mins in a thawing bath. The change in length of the samples was measured after each 50 cycles.

Compared to the reference concrete, 5 of the recycled concretes have the same high frost resistance as the reference concrete, 1 recycled concrete (mix 5) has a medium frost resistance and both recycled concretes prepared with materials from MRV have a lower frost resistance.

The petrographic examination of the hardened concrete has been carried out by means of fluorescence and polarizing microscopy on 5 thin sections of each type of concrete.

Structurally, all the concretes compared can be regarded as identical, if the structure of the aggregates is disregarded. Compared to the demands which are normally made on pavement concrete in an aggressive environment, the analyses carried out proved all samples to be acceptable.

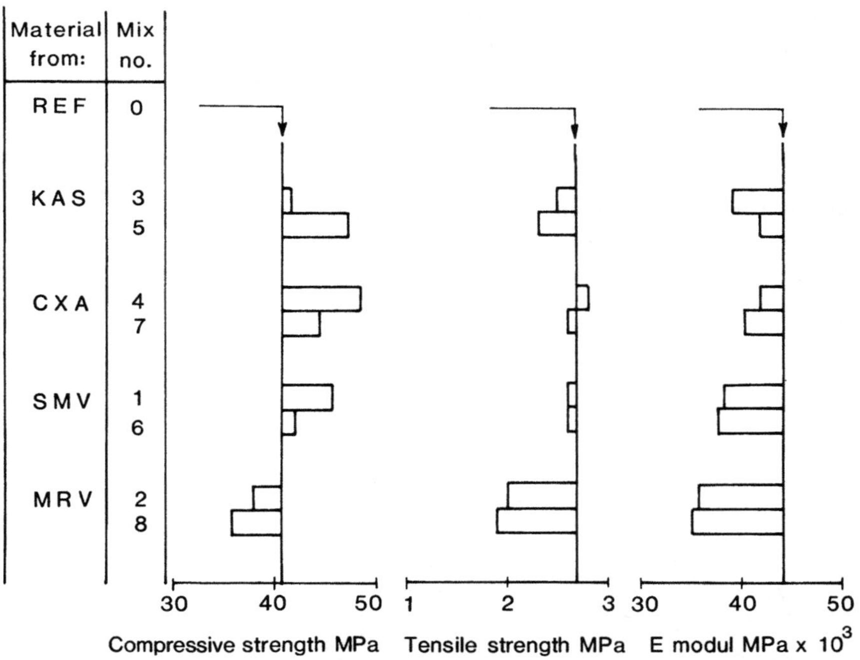

Fig. 2: Comparison of compressive strength, tensile strength and modulus of elasticity in relation to reference concrete

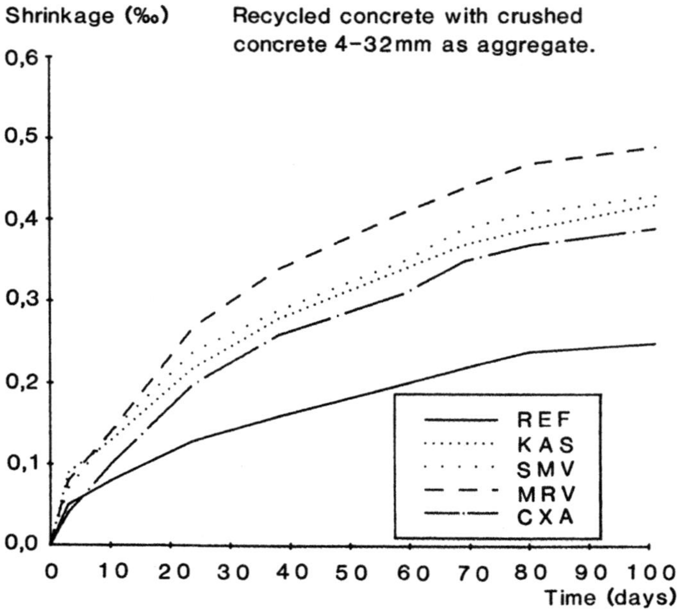

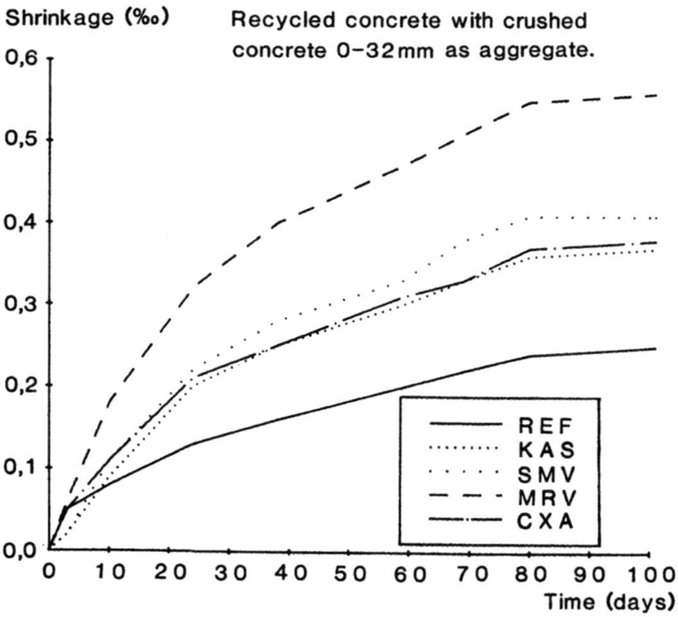

Fig. 3: Drying shrinkage

6. Conclusion

The examinations performed have shown that it is possible to recycle materials from existing concrete pavements to produce a high quality concrete for new concrete pavements, provided that the existing pavement has been made of normal concrete quality for road pavements.

This will typically be the case for concrete pavements, which are worn out due to fatigue or overloading after many years of use; this presumably applies to the greatest part of existing pavements in Denmark.

In cases where an existing pavement has shown a rapid deterioration due to alkali silica reactions and/or bad concrete quality, such as is the case for the concrete pavement of the Motor Ringroad, it is not advisable to recycle the materials when producing a new concrete pavement, unless suitability can be documented.

Crushed materials from concrete pavements, where alkali silica reactions have been proved, can possibly be recycled in such cases where a large part of the reactive material has been used up in connection with the reactions. However, examinations must first be made as to how such materials should be evaluated and which test methods should be used.

The study has also shown that the frost resistance of recycled concrete can be affected, if the mortar of the original concrete was not frost resistant. Test methods and evaluation criteria regarding this condition must also be examined more closely.

The recycled concrete should in general fulfil the same demands for quality as those made for pavement concrete with natural materials. It is important to note that the drying shrinkage for recycled concrete typically will be 50-100% higher than for concrete produced with natural materials. Also the density will be lower due to the lower density of the aggregate.

Due to the greater water absorption of these materials compared to natural materials, the mixing technique should be such as to provide an opportunity for the materials to become water saturated during the mixing process.

The part of the crushed materials which is smaller than 4 mm can be used for recycled concrete, but this is normally not advised because this part is small and needs to be supplemented under all circumstances with natural sand. Recycled sand creates problems, due to an increased water requirement in the concrete, a tendency of sticking together during storage and difficulties for determining the water absorption.

Determination of the water absorption of recycled materials should not be carried out in accordance with the test method stated in ISO. An alternative method has been examined. In this method the connection between time and the evaporated amount of water is continuously registered and based on this desorption curve the absorption is determined graphically. It is recommended that this method of testing is verified further, as regards to establishing a possible standard.

Other properties of the recycled material can be determined on the basis of normally used test methods for natural materials.

The concrete properties of the recycled concrete which were examined have been compared to a reference concrete with natural aggregates. All concretes were prepared with the same content of cement, microsilica, water and additives and can be characterised as follows:

	Concrete property	Recycled concrete compared with reference concrete
Fresh concrete	Slump Air content Density	Unchanged Unchanged Lower
Hardened concrete	Density Compressive strength Tensile strength E-modulus (dynamic) Air void system Microstructure Macrostructure Frost resistance Alkali resistance Drying shrinkage	Lower Unchanged/higher Unchanged/higher Lower Unchanged Unchanged Unchanged Unchanged *) Unchanged *) 50-100% greater

*) For recycled concrete prepared with materials obtained from the Motor Ringroad a lower resistance has been observed.

References:

[1] Knud Puckman, Anders Henrichsen, Ole Rud Hansen. "Reuse of concrete pavements, National Road Laboratory, Denmark, report 63E, 1988.

[2] Hansen T.C. Recycled aggregates and recycled aggregate Concrete. Second State-of-the-Art Report Developments 1978-1984". Rilem TC-37-DRC. January 1985, and revision March 1986 named "Developments 1945-85".

THE INSTANCES OF CONCRETE RECYCLED FOR BASE COURSE MATERIAL IN JAPAN

T. YOSHIKANE
Research Laboratory, Taiyu Kensetsu Co., Ltd.

Abstract
It has been estimated that the amount of waste-cement concrete material generated yearly in Japan reaches more than 10 million tons, amounting to several percent if the quantity of concrete laid in one year, that is, one hundred and several tens of million m^3. As the amount generated is enormous, disposal by abandoning and dumping at land reclamation sites is falling into such a state that it is not easy in regards to preserving the environment.

Under this background, Taiyu Kensetsu Co., Ltd. has made advances in the recycling of waste-cement concrete material for use as base course material and its practical use by installing a stationary plant in one of the suburbs of Nagoya City around 1976, which was the first one of its kind in Japan, and now, it has accumulated a great many achievements concerning production and construction. Therefore, the circumstances of the recycling of waste-concrete material as base course material in Japan and its features are described, and the process of its manufacture, kinds and quality characteristics of the recycled base course material and the results of the follow-up investigation after construction are reported based on the data of the achievements.
Key words: Waste matters, Recycling, Cement concrete, Concrete recycling, Pavement, Base course material, Stabilized material.

1. Circumstance and present status of recycling and utilization of waste-concrete material

The amount of waste-concrete material generated in Japan has not been accurately grasped, but at present, it is thought to have reached more than 10 million tons yearly and this estimation is on the low side. In Japan, with its limited amount of land, concrete waste cannot easily be abandoned in view of environment preservation; therefore, its recycling and utilization, and also from the aspect of effective utilization of resources, much expectation is being placed on technical development and the putting of the waste to practical use.

Under this background, a plant for the recycling and utilization of waste-concrete material was constructed around 1975, for practical uses with the recycled material being supplied as base course material. Its demand gradually expanded, and in 1984, based on those achievements, when the "Tecnical Guideline for Recycling and Utilizing Waste Pavement Materials (draft)" was established by the Japan Road Association, the way to its utilization for public construction works was opened, and this development led to increase demand thereafter. According to the investigation by the Japan Road Constractors Association in 1986, it was estimated that the amount of waste-concrete material which was used as recycled base course material, was about 130,000 t/y. Since the reuse in other fields is still very small, the material being reused is only a small fraction in view of the total amount generated, but there is great expectations seen for advancing its practical use with an increase in demand foreseen.

In addition, as for the waste-concrete material used for the production of the recycled base course material, mainly concrete without any reinforcement is the object, such as pavement concrete, the waste from secondary concrete products attached to pavements and so on, of which the waste rarely includes reinforcing bars and other admixtures, the process of recycling is simple, and the equipment and manufacture costs are comparatively low.

2. Quality standard for regenerated base course material

The quality standard for the recycled base course material using waste-cement concrete material in Japan is shown in the "Technical Guideline for Recycling and Utilizing Waste Pavement Materials (draft)" mentioned before. Fundamentally, the same standard values as those for the base course materials made by using new materials are applied. (However, the Los Angeles abrasion test and the stability test using sodium sulfate have been omitted.) As regards the kinds of recycled base course material, granular material and stabilizing-treated material using binders such as cement are the most common.

Furthermore, when waste-concrete material is taken from road pavements, the case where asphalt-concrete is overlaid on cement concrete slabs is not uncommon; therefore, it is unavoidable that asphalt-concrete waste in mixed into waste-cement concrete material. In the granular base course material, when the depth of the layer in which the recycled material is used is shallow from a pavement surface, and the ratio of mixing of asphalt-concrete waste into the recycled base course material is high, the bearing capacity of a base course is lowered due to the heat transmitted from the pavement surface, and as a result, the structural strength of a pavement as a whole is lowered; therefore, when a base course is constructed at the position where temperature reaches over 40°C, it is recommended to increase the value of modified CBR by about 10 points. Generally, when the ratio of mixing of asphalt-concrete waste in the recycled granular concrete base course material is less than 30%, it has been considered that such consideration is scarcely necessary. However, as to the material to which the stabilizing treatment using binders is applied, the effect of temperature is regarded as small, accordingly, such a restriction is not provided.

3. Kinds and quality characteristics of recycled base course material

As to the recycled concrete base course material, the typical materials that have been experimented with or put to practical use by the author are shown in Table 1. In Table 1, ① ~ ③ are those used as granular materials, but also in the case of ①, according to the investigation by opencut after three years, as shown in Photo. 1, it has been confirmed that hardening occurred due to the potential hydraulicity of the fine powder of cement hydrate contained in the recycled base course material, and the unconfined compressive strength reached 40 kgf/cm^2. For the long term, it is possible to evaluate as the base course material with stabilizing treatment rather than the ordinary granular material, and this fact is to have increased the structural strength of pavement by the increment of strength due to hardening. Also as for ③, such an effect can be expected, but because the quantity of fly-ash addition is great, it does not result in exceeding the limit of the granular material. When the bearing-capacity characteristics of the granular material were observed, these are as shown in Fig. 1 and all sufficiently exceeded the standard values mentioned before.

Table 1 Combination of materials regarding kinds of recycled concrete base course material.

Kind	No.	Material					
		Aggregate		Admixture or Binder			
		Waste-concrete	Waste base course material	Portland or blended cement	Fly-ash (waste)	Slag powder (sub product)	Waste sludge (waste) ※
Granular base course material	①	○	−	−	−	−	−
	②	○	○	−	−	−	−
	③	○	−	−	○	−	−
Stabilized base course material	④	○	−	○	−	−	−
	⑤	○	−	○	○	−	−
	⑥	○	−	−	○	−	−
	⑦	○	−	−	○	−	○
	⑧	○	−	−	−	○	−
	⑨	○	−	−	−	−	○
	⑩	○	−	−	−	○	−

※ Waste sludge of ready-mixed concrete factory.

Photo. 1 Cut sample of recycled granular concrete base course material in opencut investigation 3 years after construction.

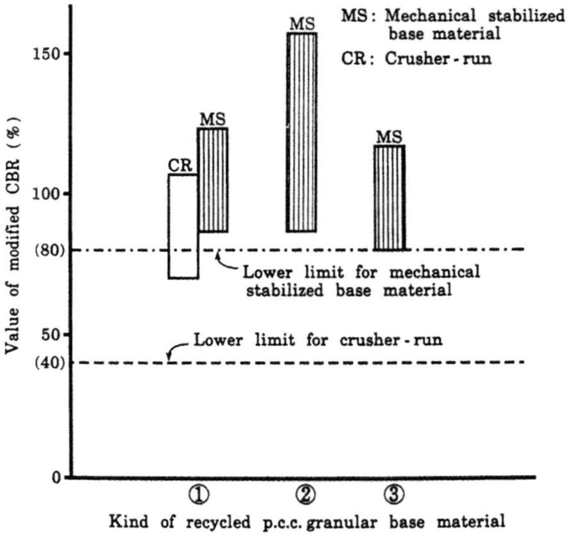

Fig. 1 Values of modified CBR for recycled granular concrete base course material.

In addition, ④ ~ ⑩ are those which were made into the base :ourse material with stabilizing treatment using various binders. As :o the binders, blast-furnace slag powder is as a matter of course a processed by-product. In addition, fly-ash and waste sludge from ready-mixed concrete factories are both industrial waste. It has been confirmed that by using these singly or in combination, the hydraulicity is demonstrated, and the excellent bearing capacity for base course increases. In this way, even waste-concrete material, in which the hydration of cement is considerably advanced due to the lapse of many years after the construction of the concrete structures, and the hydraulicity does not remaining to any great extent, becomes the hydraulic hardened body sufficiently exceeding the bearing capacity required for base course materrial, by making cement hydrate into fine powder by pulverizing and the appearance of fresh-fracture surfaces, as the hydration reaction is stimulated again by the calcic binders of low grade and fly-ash. In Fig. 2, the relation between the age of these stabilizing treatment mixtures and the compressive strength is shown.

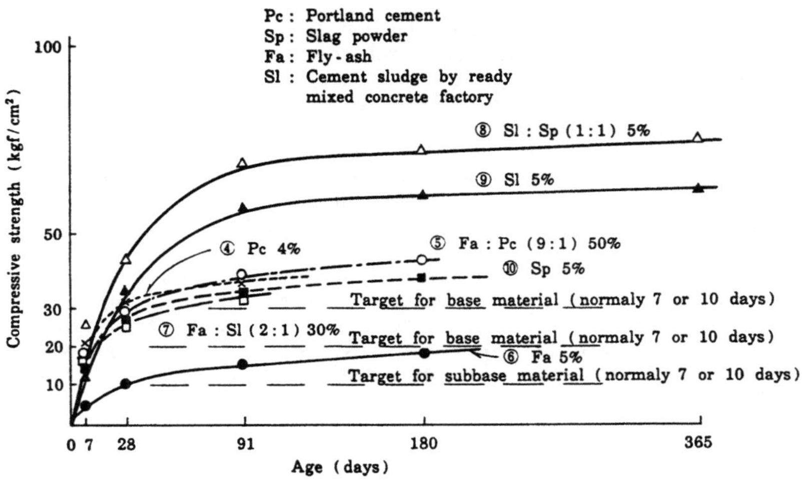

Fig. 2 Relation between age and unconfined compressive strength in recycled base course material with a stabilizing treatment.

In all the stabilizing treatment mixtures, the diversified results were obtained in the state of strength manifestation according to the kinds of binders and the ratio of mixing, but those that are typical in the range of practical use including the economical efficiency are shown in Fig. 2. Though there is some difference among respective materials, it may be considered that the time to nearly reach the upper limit of compressive strength is after one to three months, and it is known that those shown in the figure can all sufficiently cope with the standard values given to the base course material with stabilizing treatment in Japan. When the case of using other binders than Portland cement was compared with the case of using Portrand cement, the strength at the age from 7 days to 28 days arose somewhat low, but in the manifestation of strength in the long term, the rate of increase exceeds that in the case of using Portland cement. It is considered that when the practical use is made by understanding these features, there is not any hindrance. Among others, it was noteworthy that with the combination of waste-concrete material and the waste sludge from ready-mixed concrete factories of ⑨, the same strength as in the case of Portland cement addition was obtained. In addition, also that being added with the fly-ash of low quality to them is effective for practical use.

4. Regeneration of waste-concrete material and examples of construction with recycled base course material

An example of the manufacture of recycled base course material from waste-concrete material is shown in Fig. 3 with a flow chart, and the whole view, the stockyard for waste-concrete material and a primary crusher of this plant and the recycled concrete base course material (crushed stones for mechanical stabilization) are shown in Photos. 2 ~ 4. The feature of this plant is to use an impact crusher as a primary crusher in order to remove, as far as possible, mortar from aggregate surfaces. The production capacity is 100 t/h, and the operation was started in 1976. As regards the production progress, the achievement of 68,000 t in 1985, 97,800 t in 1986, and 90,500 t in 1987 was attained, and as a plant for recycled concrete base course material, it has the largest recycling capacity in Japan. The quality-control data in this plant are shown in Table 2, and the execution control data in pavement works are shown in Table 3. From these results, it is not considered that there is particular difference as compared with the control data in the production and execution of ordinary crushed stone base course material.

Photo. 2 Recycling plant and stockyard for waste-concrete material.

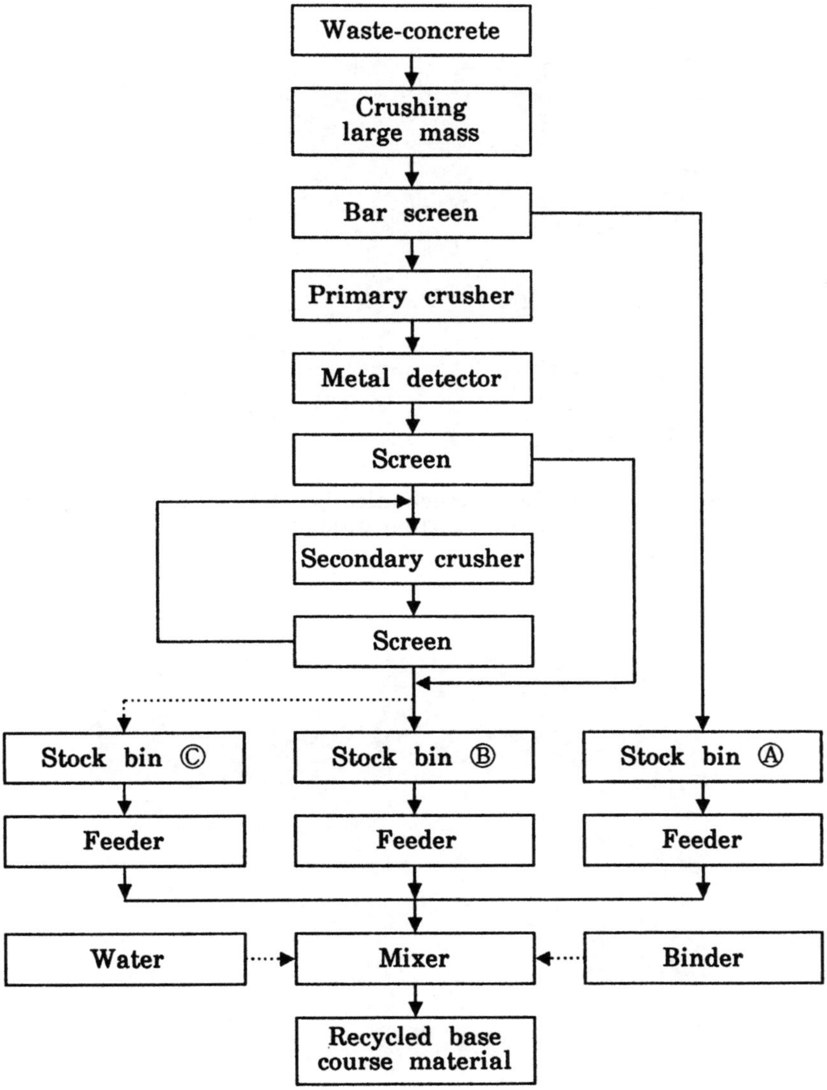

Fig. 3 Flow chart for production of recycled base course material.

Photo. 3 Recycled mechanically stabilized base course material during unloading in field.

Photo. 4 Mechanically stabilized base course in field.

Table 2 Example of quality-control data in recycled concrete base course material (mechanically stabilized base course material) production plant (January~December, 1986).

	n	Mean	Standard deviation	Max.	Min.	Specification
Maximum dry density (g/cm^3)	24	1.969	0.022	2.002	1.894	–
Optimum moisture content (%)	24	9.81	0.42	11.00	9.25	–
Modified CBR (%)	24	86.1	5.5	97.5	80.0	more than 80
PI	24	N.P	–	–	–	less than 4

Table 3 Example of gradation analysis in recycled concrete base course material (mechanically stabilized base course material) production plant (January~December, 1986).

	Weight percentage passing sieve (%)					
Sieve opening (mm)	n	Mean	Standard deviation	Max.	Min.	Specification (Mechanical stabilization)
50	12	100.0	–	–	–	100
40	12	99.8	1.6	100.0	98.1	95–100
20	12	75.4	3.2	79.0	70.4	60–90
5	12	37.9	2.3	40.1	32.6	30–65
2.5	12	25.8	2.3	28.6	22.6	20–50
0.4	12	13.2	2.2	15.8	10.3	10–30
0.074	12	3.3	0.6	3.9	2.2	2–10

Table 4 Example of field-density test results for base course using recycled concrete base course material.

	Recycled concrete for crusher-run base course		Recycled concrete for mechanically stabilized base course	
	Degree of compaction (%)		Degree of compaction (%)	
	Lower layer	Upper layer	Lower layer	Upper layer
n	25	25	9	11
Mean	97.4	98.7	98.6	98.6
Standard deviation	0.63	0.38	0.23	0.48
Max.	98.4	99.4	99.1	99.3
Min.	96.0	98.7	98.3	98.1

5. Future Perspective

The circumstance of the recycling and utilization of waste-concrete material as base course material has been mentioned concerning the achievement, the outline of the present recycling techniques and the kinds and properties of the recycled base course materials, and the cases of the manufacture of and construction with the recycled base course materials. As mentioned, it was found that with the recycled materials of concrete used as base course material, that strength of base course became remarkably high over the long term due to hardening and integration due to the advance of the hydration of remaining unhydrated material if a small quantity of water existed, without the use of binders. Also, the structural strength of pavements became high as a result of this.

Generally, when pavement is used for traffic after construction, the lowering of bearing capacity gradually advances due to the fatigue by cyclic loading in the materials composing layers and hardened substances of the pavements, as the structural strength gradually decreases, and the residual life of the pavement is decreased.

However, in the case of using the recycled concrete base course material, as the structural strength of pavement increases for a considerable time after putting the pavement to use, because the material is the property of increasing the strength for a long period after construction, they are to do the action for extending the life of pavements.

This tendency is the phenomenon observed also in the case of recycling with stabilizing treatment, and in this way, there are features in the recycled concrete base course materials, that cannot be observed in conventional base course materials, therefore, the author would like to emphasize that the recycled concrete base course materials are those to be utilized to a great extent by making the best use of such advantages hereafter.

References

Japan Road Association, Tecnical Guideline for Recycling and Utilizing Waste Pavement Materials (draft), (1984).

Clean Japan Center (1982) Recycling Technique (Construction waste).

Yoshikane T. (1981) A suggestion of recycling for waste in ready-mixed concrete factory, All Nippon Ready-mixed Concrete Union, pp 149－154.

Funabashi M. and Miyata T. (1979) Stabilization for waste-cement concrete by sludge, Proceeding of XVI Japan Road Congress, Japan Road Association, pp 267－268.

CRUSHED CONCRETE USED AS BASECOURSE MATERIAL ON RUNWAY 04R-22L AT COPENHAGEN AIRPORT

CHRISTIAN BUSCH
Airports and Pavement Department, Cowiconsult

Abstract
Reconstruction of runway 04R-22L at Copenhagen airport, Kastrup was undertaken during the summer of 1983, reusing the materials of both the original PCC pavement and the subsequently added asphalt overlays. The PCC was reused in an unbound basecourse layer. This application of the demolished concrete was chosen after initial tests and studies during the design phase, confirming that the PCC could be processed to fulfil the gradation requirements of a basecourse material. The tests also demonstrated that the resultant product could be compacted to form a layer with higher bearing capacity than traditional gravel material. The reconstruction of the 3.3 km long runway was executed according to the plan in less than 5 months. Quality control of the crushed concrete basecourse included nuclear probe densimeter measurements as well as plate bearing tests. It was found that the crushed concrete could be compacted at high water contents, hereby providing a material that is very useful when time schedules are important in adverse weather conditions.
Key words: Airport pavements, PCC demolition, Falling plate, PCC crushing, Test section, Pavement design, Quality control, Asphalt recycling.

1. Construction history of runway 04R-22L

1.1 From grass field to PCC pavements
Copenhagen's International Airport on the island of Amager dates back to 1925, when operations were initiated on a grass field. In 1941, it's 3 runways, arranged in the traditional triangular pattern required by the more crosswind-sensitive aircraft of that era, were reconstructed as undoweled 200 mm PCC pavements on a subbase of uniformly graded marine sand. During the years to follow, numerous changes were made to the airport's layout, but the runway 04-22 remained in service and was gradually lengthened to 3300 m, with the additional sections also being constructed as PCC pavements and thicknesses up to 300 mm.

1.2 Cracking and Asphalt overlaying
The introduction of still heavier aircraft led to cracking of slabs and pumping of material at the undoweled joints. Several asphalt overlays were applied to remedy this deficiency, and the total

asphalt thickness eventually reached approximately 300 mm on the central part of the runway. Differential slab movements and reflective cracking continued, however, to be a major problem.

By the end of the sixties, traffic volume made the construction of a parallel runway, 04L-22R, necessary. With this project came the first real opportunity for a reconstruction of the distressed runway, now renamed 04R-22L. This opportunity was, however, immediately lost with the imposing of a moratorium for all major construction works in Kastrup Airport, since it had been politically decided that a new main air terminal for Copenhagen was to be constructed on the island of Saltholm. This island is so close to Amager that operations at Kastrup would have to be terminated.

In the following decade, economical as well as environmental considerations led to successive postponements of the Saltholm project, until it was finally cancelled in 1980. By that time it had become apparent that a rehabilitation of runway 04R-22L was required as soon as possible.

2. Reconstruction alternatives

2.1 Necessity for reuse

From the onset of the rehabilitation planning, the logistics of the project dictated a solution that involved some sort of material reuse. A project, based exclusively on new materials would mean discarding approximately 75,000 tonnes of old asphalt material and 150,000 tonnes of PCC. The disposal of such huge amounts of debris, and the transport of replacement materials would involve moving some 0.5 million tonnes, of which a considerable amount would have to be taken by truck through Copenhagen.

2.2 Test programme

During the summer of 1982, a comprehensive test progamme was initiated, including laboratory tests on asphalt and PCC core samples, as well as the construction of a 100 m long full scale test section. On this was determined the practical requirements for a satisfactory compaction of a crushed concrete basecourse, and a number of experiments concerning flyash-stabilization of the uniformly graded sand subbase were carried out.

It was deemed that the benefits (if any) reaped from stabilization did not make up for the complicated process of mixing the flyash thoroughly with the subbase material. When it was furthermore noted that the satisfactory compaction of the crushed concrete basecourse was independent of subbase stabilization, this aspect of the project was discarded.

One very important result from the full scale test was the determination (based on quasi-static plate loading tests) of design values for the unbound materials. The values are summarized in table 1 below.

Table 1. Unbound material design values

Material description	E-modulus (MPa)
Crushed concrete basecourse (min. 25 cm thick)	400
Sand subbase (upper 25 cm of a 50 cm layer)	200
Sand subbase (lower 25 cm of a 50 cm layer)	100
Subgrade	25

3.3 Pavement design

Pavement design was based on the analytical method, relating calculated resilient strains to predicted pavement performance. The criteria adopted were based on the work of Kirk (1972) for asphalt materials, and Shell (1978) for unbound materials.

The E-modulus of the Crushed concrete basecourse was 33 % higher than the value normally found for Danish gravel basecourse material, hereby reducing strains in the bottom of the asphalt basecourse. This permitted asphalt layers to be designed 10 to 12 % thinner than would otherwise have been the case. The reason for this increase in E-modulus is assumed to be the particular structure of the crushed material. With the larger aggregates mostly consisting of a hard nucleus of stones from the original concrete, covered with a softer crust, the number of contact points between the base course material's particles is higher than for traditional gravel. This number of contact points is one of the factors governing unbound material stiffness. One factor that should not be overlooked is, on the other hand, the corresponding increased brittleness, which tends to increase material degradation during compaction, but will probably not present any problem, given the stresses under normal traffic conditions. One final point that has not been fully clarified is the possibility that some of the original cement content may become reactive, hereby creating a hardening effect. It would be desirable to carry out a long-term surveillance project to investigate this possibility.

The asphalt tests confirmed that the existing materials were suitable for production of an asphalt basecourse which could contain a very high content of recycled material, before it became inferior to one produced from virgin materials.

As for utilization of the crushed concrete as aggregate in a new PCC pavement structure, ongoing research at that time - later reported by Hansen and Narud (1983) - pointed out that both strength and deformation characteristics would be inferior to those of PCC made with new aggregates. Since the durability aspect was also undecided, the possibility of a PCC pavement based on reuse was disregarded.

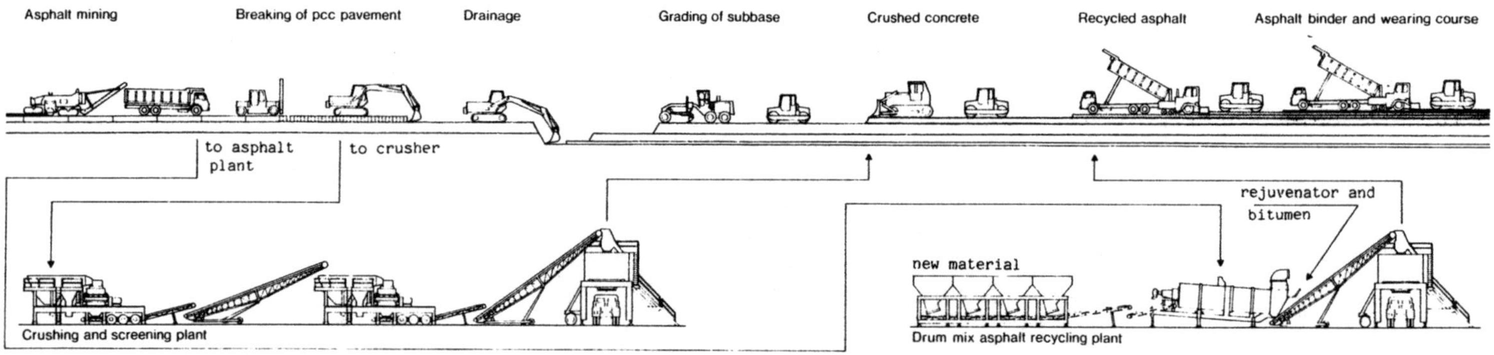

Fig. 1. Recycling sequence with concrete processing to the left and asphalt processing to the right

During the preliminary design phase, the three alternatives considered were therefore:

(a) Replacement of all cracked and decomposed sections of the existing asphalt layers and application of an additional asphalt overlay and wearing course to provide the necessary strengthening and improve the runway longitudinal profile. The old concrete pavement would be retained and drainage along the edges provided to improve the bearing capacity of the sand subbase.
Estimated cost was 8 million US $.

(b) Construction of a flexible pavement, consisting of an aggregate base course produced from crushed concrete, and a bituminous base course produced with a maximum of recycled material - an amount between 60 % and 65 % was found to be possible. The pavement would be topped with high-quality binder- and wearing courses, and the total reconstruction scheme would make a more radical improvement of the drainage possible.
Estimated cost was 8 million US $.

(c) Construction of a rigid pavement, consisting of plain PCC slabs made from new materials, placed on a foundation with an asphalt and a crushed concrete basecourse, both made solely from existing materials. This scheme would also allow a thorough improvement of the drainage system.
Estimated cost was 9 million US $.

For the final preparation of tender documents, alternative (a) was disregarded, mainly because it wasn't seen as an improvement over the patchings carried out previously, and didn't offer any degree of certainty that the basic deficiencies of the runway would be overcome.

3. Tender and construction

3.1 International tender

The project was put out for international tender, with several internationally recognized contractors offering bids. The contract, however, was won by a cartel of the four major danish asphalt contractors, and the cheapest alternative was, as expected, the flexible pavement solution.

To do the job, they set up a newly purchased 300 tonne/hr Marini asphalt recycling plant, and a three-stage crushing plant with a jaw crusher and two cone crushers.

3.2 Reuse material processing

The construction process, shown schematically in fig.1, began with mining of the old asphalt by cold-milling, whereupon the PCC slabs were broken up by falling plate machines, hitting the slabs for every 50 cm. Two machines were commissioned, reaching an average daily demolition rate of 8,000 m² per machine. Removal of the broken concrete slabs proved to be the most difficult operation, since it was impossible to avoid digging up some 10 % of the uniformly graded subbase sand material. The effect of this sand content is clearly seen in fig.2 as an irregularity on the otherwise ideally graded material.

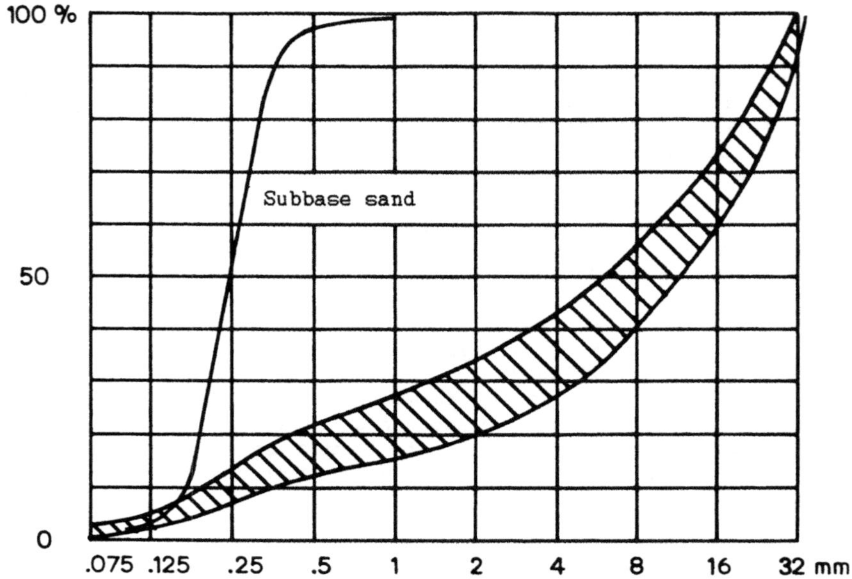

Fig. 2. Sieve analysis of crushed concrete with sand content

3.3 Crushed concrete basecourse laying and compaction

After completion of the drainage works, and subbase surface adjustments to rectify the longitudinal profile, the crushed concrete basecourse was placed in a two-stage process. On the relatively unstable subbase surface, the first lift of basecourse material was spread by dozers, keeping off the subbase surface. After compaction of this lower basecourse layer, the upper lift was placed by means of asphalt pavers, which guaranteed an extremely accurate final level of the unbound basecourse layer. The width of the paving lanes had to be limited to 5 meters, since there was a tendency to seperation of the material since stones are more easily transported than fines by the snail transporters of an asphalt paver. The compaction itself led to some degradation of the basecourse material, but the close resemblance to a Fuller-curve was retained, as shown by fig.3.
Compaction control for the lower layer was made only in terms of number of roller passes. On the upper layer, traditional density measurements by means of sand filling resp. nuclear densimeter measurements were supplemented by plate bearing tests to check whether the design prerequisites were met. If all materials were up to standard, the resultant surface modulus should be higher than 120 MPa. Where this was not the case, additional compaction was applied, always resulting in acceptable surface moduli, as demonstrated in fig. 4, which shows the surface moduli measured on two dates before, and one after additional compaction was applied.

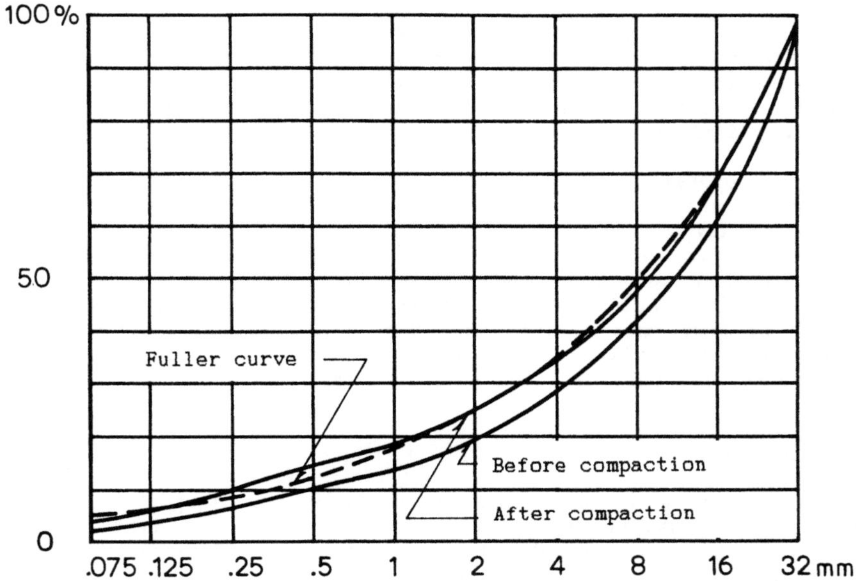

Fig. 3. Degradation of crushed concrete under compaction

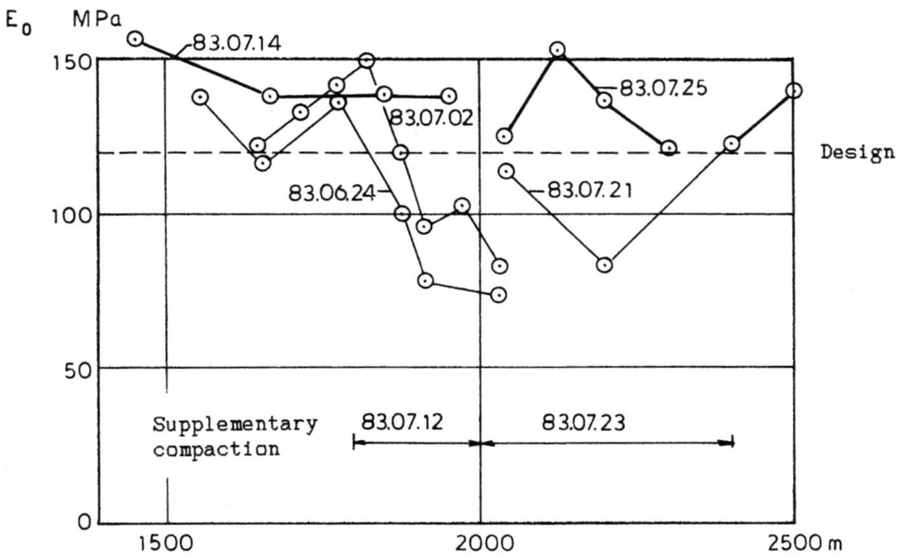

Fig. 4. Compaction control by plate loading tests and subsequent supplementary compaction effort. Numbers indicate date of activities

The admixture of water, which was often required during the extraordinary warm summer presented no problems. The fines of the crushed concrete material are non-plastic, so that over-watering is virtually impossible - rollers were seen working on water-logged surfaces without any difficulties.

3.4 Airport operation
The planned duration of the project was four months and three weeks, starting on May 9. Advantage was taken of the "light" Scandinavian summer nights to work two shifts, so that this schedule was in fact met. Moreover, due to careful planning of temporary aircraft approach control facilities, the runway was only non-operational for a period of four months, and with excellent cooperation between all parties, notably airline pilots and air traffic controllers, flights were handled with only occasional minor delays.

4. Conclusions

The reconstruction of runway 04R-22L at Copenhagen airport, Kastrup, was in all respects an unqualified succes. It was the first project of its kind on such a grand scale in Denmark, and remarkable in the respect that both asphalt and concrete was reused. From it may be drawn the following conclusions:

- More detailed pavement and materials investigations are needed than with a traditional project;
- Special equipment and techniques are required to demolish and extract concrete for use in production of a high-quality unbound base material;
- Special crushing equipment with extensive possibilities of process adjustments is needed to produce a high-quality unbound base material from demolished concrete;
- If asphalt pavers are used to place the crushed concrete base-course material, special care should be taken to prevent separation of the material. Compaction is otherwise fairly uncomplicated, and the material is very insensitive to too high water content, since the fines are non-plastic;
- The unbound base material produced from demolished concrete was found to be superior to traditional Danish gravel pit materials, with a 33 % higher E-modulus, hereby allowing a thickness reduction of the asphalt layers;
- It is possible to execute a reuse project on a busy airport without seriously affecting the flight operations, provided that an off-season period is chosen, and that high priority is put on preparatory planning as well as coordination and information activities through the project period;
- Reuse improves the economy. The savings realized on the actual project are estimated to 20-40 %, when compared to traditional construction methods, and environmental impact was drastically reduced, with reduced truck traffic, no dumping of demolished materials and saving of gravel resources for other purposes.

References

Kirk, J.M. (1972) Relations between Mix Design and Fatigue properties of Asphaltic Concrete. Third International Conference on the Structural Design of Asphalt Pavements. University of Michigan, pp 241-247.

Shell (1978) Shell pavement design manual - asphalt pavements and overlays for road traffic. Shell International Petroleum Company Limited, London.

Hansen, T.C. and Narud H. (1983) Strength of recycled concrete made from crushed concrete coarse aggregate. Concrete International Vol.5, No.1 pp 79-83.

Bejder, J. (1983) Copenhagen - Kastrup's Recycled Runway. Airport Forum No.6/1983. Bauverlag GmbH, Wiesbaden.

Fjellerup, F.E. (1983) Genbrug af beton i Kastrup Lufthavn. Demolering og genbrug af beton. Dansk Betonforening, Publikation nr.20.

For Product Safety Concerns and Information please contact our EU representative GPSR@taylorandfrancis.com Taylor & Francis Verlag GmbH, Kaufingerstraße 24, 80331 München, Germany

Batch number: 10725973

Printed by Printforce, the Netherlands